Salmonid Reproduction

An International Symposium

Review Papers

Bellevue, Washington • October 31–November 2, 1983

Robert N. Iwamoto & Stacia Sower, Editors

Washington Sea Grant Program
University of Washington • Seattle

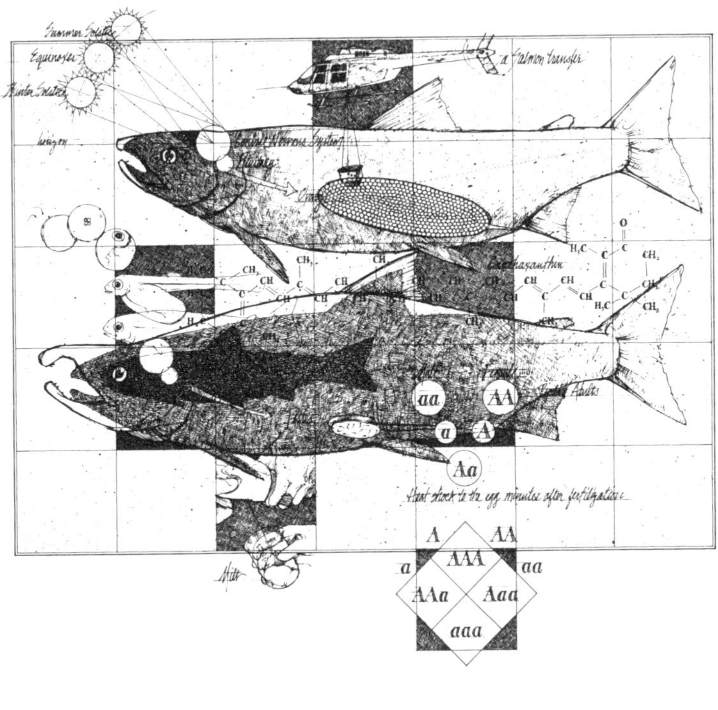

First published in 1985 by
Washington Sea Grant Program
University of Washington

Printed in the United States of America

Support for publication of these proceedings was provided in part by grant number NA81AA-D-00030, projects R/A-36 and A/PC-5, from the National Oceanic and Atmospheric Administration to the Washington Sea Grant Program.

Key Words: 1. Salmonid reproduction 2. Salmonid endocrinology 3. Salmonid genetics 4. Salmonid nutrition 5. Environmental aspects of salmonid reproduction 6. Salmonid husbandry 7. Salmonid broodstock management 8. Proceedings

LIBRARY OF CONGRESS CATALOGING IN PUBLICATION DATA
Main entry under title:
Papers presented at the International Symposium on Salmonid Reproduction.
Includes bibliographies.
1. Salmon–Reproduction–Congresses. 2. Fish-culture–Congresses. I. Iwamoto, Robert N., 1946–
II. Sower, Stacia, 1950– III. Washington Sea Grant Program. IV. International Symposium on Salmonid Reproduction (1983 : Bellevue, Wash.)
SH167.S17S28 1985 639.3'755 85-13764
ISBN 0-934539-00-6 (pbk.)

Washington Sea Grant Program
College of Ocean & Fishery Sciences
University of Washington 98195

WSG-WO 85-2
June 1985
$ 10.00

Contents

Symposium Sponsors

Washington Sea Grant
University of Washington

with additional financial support from

Pacific Sea Grant Colleges
Oregon Sea Grant College
Hawaii Sea Grant College
California Sea Grant College
Alaska Sea Grant College

National Marine Fisheries Service

U.S. Fish & Wildlife Service

Steering Committee

General Chairmen
William K. Hershberger
Robert N. Iwamoto
School of Fisheries
University of Washington
Seattle, WA

Technical Program
Robert N. Iwamoto
School of Fisheries
University of Washington
Seattle, WA

Stacia Sower
Department of Zoology
University of New Hampshire
Durham, NH

Coordination & Publicity
Patricia A. Peyton
Communication Services
Washington Sea Grant Program
Seattle, WA

Local Arrangements
Terry Y. Nosho
Marine Advisory Services
Washington Sea Grant Program
Seattle, WA

Finance
Conrad Mahnken
Northwest & Alaska Fisheries Center
National Marine Fisheries Service
Seattle, WA

Proceedings
For Aquaculture
Colin Nash
Kramer, Chin & Mayo
Seattle, WA

For Washington Sea Grant
Patricia A. Peyton
Communication Services
Washington Sea Grant Program
Seattle, WA

Advisory Committee

Brian Allee
Prince William Sound Regional Aquaculture Corporation
Cordova, AK

Edward Brauer
Domsea Farms, Inc.
Bremerton, WA

Walton Dickhoff
University of Washington
Seattle, WA

Edward Donaldson
Fisheries & Oceans Canada
West Vancouver, BC

Lauren Donaldson
University of Washington
Seattle, WA

Graham Gall
University of California
Davis, CA

Aubrey Gorbman
University of Washington
Seattle, WA

Charles W. "Bill" Hopley
Washington Department of Fisheries
Olympia, WA

Harold Kincaid
U.S. Fish & Wildlife Service
Kearneysville, WV

James Lannan
Oregon State University
Newport, OR

Phil Mackey
Mt. Lassen Trout Farm
Red Bluff, CA

Carl Schreck
Oregon State University
Corvallis, OR

ACKNOWLEDGMENTS

The International Symposium on Salmonid Reproduction evolved from two prior workshops held in Seattle, Washington (in 1980 and 1981), sponsored by the Washington Sea Grant Program and coordinated by Terry Nosho, Sea Grant aquaculture specialist. Though originally intended for a regional audience, the need for an international forum to promulgate information exchange and to discuss pertinent and current issues on salmonid reproduction became apparent. Escalation to that level of organization required the cooperation of many individuals and institutions whose contributions it is now our pleasure to acknowledge.

We are indebted to the Washington Sea Grant program for its enthusiastic support and willing sponsorship of the symposium from conception to completion. In particular, we would like to acknowledge Patricia Peyton, a member of the steering committee, and her staff for their excellent organization and expertise with symposium communications. The efforts of William R. Davis, Terry Nosho, Louis Echols, Alan Krekel, and Scott Johnson are also gratefully recognized. The illustration that has been used to identify this symposium was done by Kirk Johnson.

We would like to acknowledge other members of the steering committee–William Hershberger, Conrad Mahnken, and Colin Nash, for their organizational contributions. Special thanks go to Carl Schreck and Edward Donaldson of the advisory committee who provided valuable assistance and criticisms.

Financial contributions from the Pacific Sea Grant College Programs in Alaska, California, Hawaii, and Oregon; from the National Marine Fisheries Service; and from the U.S. Fish & Wildlife Service enabled us to invite several international participants who otherwise would not have been able to attend.

Finally, we would like to thank the speakers and participants at the symposium and the contributors to this record of the first, but hopefully not last, symposium on salmonid reproduction.

Robert N. Iwamoto and Stacia A. Sower
July 1984

PROLOGUE

In a figurative sense, salmonid resources around the globe have been buffeted and shaped by multi-directional forces representing social, economic, and political interests. Recently, much concern has been expressed about the diminishing levels of many wild stocks (c.f., *Aquaculture,* 1981 and *Aquaculture,* 1983) and about the status of both public and private enhancement efforts to maintain and replenish such stocks. These varied interests have stimulated a great deal of basic and applied research on salmonid biology. Most research results have been beneficial; however, in some instances, new techniques are being made available to salmon culturists without complete information about their benefits, limitations, and cost-effectiveness.

This international symposium on salmon and trout reproduction was designed for exchange of information on one very important and sometimes neglected aspect of salmonid culture–reproduction. Questions raised as a result of an increased research effort and the growing interest of hatchery managers, sea ranchers, and net-pen culturists prompted us to offer a symposium covering five major aspects of salmonid reproduction:

- Endocrinology
- Genetics
- Nutrition
- Environmental factors
- Husbandry

Specific topics included precocious maturation, sex reversal, induced ovulation, dietary requirements of captive broodstock, genetic studies in reproduction, temperature and photoperiod effects on maturation, and the normal endocrine events preceding and concurrent with maturation and spawning.

Attended by more than 300 participants, the two-and-one-half-day symposium featured 84 individual presentations and an evening of roundtable discussions. Slightly fewer than half of the individual contributions have been published in a special issue of *Aquaculture*, volume 43 (1984). This volume, published by Washington Sea Grant, includes review papers, summaries of the roundtable discussions, and abstracts of all other papers.

Keynote Addresses

Speakers

Lauren R. Donaldson
William J. McNeil

THE CHALLENGE

Lauren R. Donaldson[1]

For millions of years the salmonid fishes have been reproducing successfully in the temperate waters of the northern hemisphere. The more than sixty species and countless number of racial stocks should be more than adequate proof that the fish "know how to do their thing."

In collection of books on early fish culture is a delightful little volume, *Practical Trout Culture*, by J.H. Slack, M.D., published in 1872, the same year the first salmon hatchery was put into operation on the McCloud River, California. Dr. Slack records the early attempts to understand the reproduction of fishes. One story is particularly intriguing. It involves a Frenchman named Joseph Rémy.

"Rémy was a fisherman who gained his livelihood by the capture of trout in the streams of the Vorges Mountains. He noticed with regret the rapid disappearance of his favorite fishes, and being, though uneducated and ignorant, active, energetic, and persevering, devoted himself for several years to the study of their habits, especially during the spawning season. The excessive drought during the summer of 1842 favored his investigations. It was, of course, impossible for one man to keep a constant eye upon a school of fishes; nature would demand rest. Rémy therefore associated with him a tavern-keeper named Gekin, who alternated with him in his observations. So earnestly were these pursued, that in one instance, during the full of the moon, a school of trout was kept constantly in view for four consecutive days and nights. The result was the discovery of the process of reproduction, which they at once put into practice. The results of their observations were kept secret until 1848 when they were reported on by a Dr. Haxo. Rémy became at once a celebrity; he was invited to Paris, and the fisherman, but a few months previous utterly unknown, was an honored guest at the table of the president of the Republic."

Here we are assembled, 141 years after Rémy made his great discovery, trying to understand the mystery of salmonid reproduction. Either Rémy's claim to fame was greatly exaggerated or salmonid reproduction is much more complicated than many had assumed.

In this symposium, we must try to take advantage, not only of "discoveries" of Rémy, but of the thousands who have worked in this area—including the papers and informative posters of this symposium—to produce more "bigger and better" fish.

[1] Professor Emeritus, School of Fisheries, University of Washington, Seattle

SALMON RECIPE[1]

William J. McNeil[2]

1 part science
2 parts technology
3 parts economics
10 parts politics[3]

The major challenge before us is to mix the ingredients to achieve synergism—not antagonism. My thesis is based on four postulates:

- *Political processes* lead to policies on resource utilization and development.
- *Economic value* of a resource is the key stimulus for policy decisions.
- *Technology* creates and enhances economic value.
- *Science* is the foundation of technology.

Most of you who are participating in this symposium are scientists and technologists. You and your work are basic to the "Salmon Recipe." You provide synergism for economic progress and policy decisions. Your symposium topic, salmonid reproduction, is a subject which is critical to the role of science and technology in the "Salmon Recipe."

Salmon can be described by three words, each of which begins with the 22nd letter of the English alphabet—the letter "V." The words: valuable, vulnerable, and variable. Salmon are *valuable* because demand exceeds supply. Competition for a limited supply is severe. Increases in supply resulting from science and technology are quickly absorbed by demand.

Salmon are *vulnerable* to man-made changes in habitat and to overfishing. The chronicle of stocks which have been depleted, decimated, and annihilated is growing. Because science and technology have provided tools to mitigate, rehabilitate, and enhance salmon, the chronicle of stocks which have been newly created and/or increased is also growing.

Salmon are *variable* due to natural changes in their environments. Natural survival can vary more than an order of magnitude. The economic and political consequences of a variable supply of salmon can be severe. A better scientific understanding of mortality processes is essential if we are to predict and potentially reduce this variability.

Salmon are complex organisms. They live in complex ecosystems. Man places great economic demands on them. Policies affecting the allocation, utilization, and development of salmon resources are debated daily within and

[1]This is not a Betty Crocker recipe.
[2]General Manager, Oregon Aqua-Foods, Inc., Springfield, Oregon
[3]Warning: improper mixing of ingredients can cause indigestion and nausea.

among institutions and political jurisdictions. Policy makers need and seek guidance from the scientific/technical community in the decision-making process. The quality of decisions is directly related to the quality of scientific/technical information and advice.

I will cite three examples to illustrate the "Salmon Recipe:"

- U.S./Canada salmon treaty
- Northwest fish conservation and power plan
- Industrial salmon ranching

The following quotation is from Issue 24 of the *Ocean Law Memo*, Oregon State University Sea Grant Program:

> A full-fledged salmon war is shaping up in the Pacific Northwest, one in which there can be no winners. Unlike salmon skirmishes of the past between commercial fishermen and Indian tribes, between ocean trollers and river gillnetters, between recreational and commercial fishermen, this simmering and potentially volatile conflict is all the more alarming because there are no courts of law that have the power to compel the parties to resolve their differences.

The United States and Canada have been attempting to negotiate a Pacific salmon treaty for nearly 20 years. Negotiators concluded and signed a draft treaty last December, but it has not been ratified. Instead, it has been returned to the negotiators.

Economic and conservation questions stimulated efforts to draft a treaty. Scientific inputs provided the foundation to draft a treaty document. The decision process is now largely a political one. Ratification will provide a political solution to the most serious salmon conservation problem affecting the entire eastern Pacific rim. Failure to ratify a workable treaty could result from improper mixing of ingredients in the "Salmon Recipe." The resulting antagonism could lead to frustration and possible political decisions having serious consequences on the salmon resources of the United States and Canada.

The Northwest fish conservation and power plan provides a vehicle to address some long-standing problems. Twenty-eight federal dams have been constructed in the Columbia basin over the last 50 years. These, along with non-federal dams, have materially reduced the salmon resources of the Columbia.

The Pacific Northwest Electric Power Planning and Conservation Act of 1980 created a Regional Power Planning Council and directed it to develop a basin-wide program to preserve and restore anadromous fish stocks.

Prior to the Power Act, fishery interests were required to shoulder the "burden of proof" on losses of anadromous fish to dams before corrective actions were taken. The Power Act now allows the fishery agencies to take a more definitive role in planning. The key requirement is that fishery recommendations be supported by "best available scientific knowledge."

Thus, knowledge is recognized in the law as the basis for decision making affecting fisheries and the use of water resources in the Columbia. The fisheries plan that has emerged is innovative and pays more attention to conservation and enhancement than previous programs. It is founded on "best available scientific knowledge." Thus, science provides synergism in this application of the "Salmon Recipe."

Industrial salmon ranching probably contributes about 25 percent of the world harvest of salmon. Hatchery production of juveniles has doubled in the last 10 years and is likely to double again in the next 10 years.

Salmon ranching is a highly visible step in a transition from a hunting to a farming economy. Problems with conservation and management of wild stocks have provided a stimulus for ranching and other forms of aquaculture. However, the legal structure governing salmon is based on hunting. The politics of transition from hunting to farming are very difficult, and severe tests are presently taking place in Oregon and elsewhere.

The political arguments between salmon ranching and traditional commercial fishing have highlighted a number of questions which present challenges to the scientific/technical community. Among the many issues are questions about the capacity of marine waters to grow salmon, about genetic effects of hatchery fish on natural populations, and about fisheries on mixed stocks of hatchery and wild fish. The transition from hunting to farming is difficult. Science and technology provide an essential foundation for progress in economic growth and policies supporting benefits to society. Every reasonable effort must be made to seek the appropriate blending of ingredients in the "Salmon Recipe" to insure an orderly transition.

From these three examples, it is evident that certain ingredients are integral parts of the "Salmon Recipe." Science produces technology. Technology generates economic activity. Economic activity triggers political decision making. Good science provides a basis for proper blending of the ingredients. It is important for fishery scientists to become involved with all elements of the recipe, including politics, to insure that the resulting product is wholesome and nutritious for society.

Endocrinology

Session Leaders

Yoshitaka Nagahama

Edward M. Donaldson

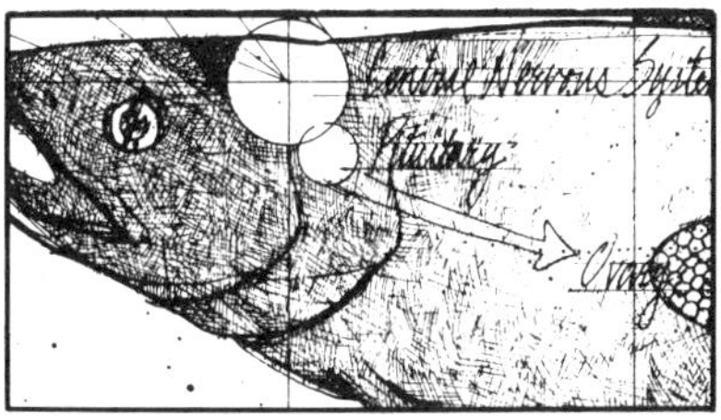

ENDOCRINE CONTROL OF FINAL GAMETE MATURATION IN SALMONIDS

Yoshitaka Nagahama,[1] Graham Young,[2] Hiroshi Ueda,[3] Hirohiko Kagawa,[3] and Shinji Adachi[1]

Abstract: Endocrine control of final gamete maturation in salmonids has been investigated in our laboratory using four species of salmonids, amago salmon (*Oncorhynchus rhodurus*), masu salmon (*O. masou*), chum salmon (*O. keta*), and rainbow trout (*Salmo gairdneri*). Our biochemical studies combined with data from the application of *in vitro* techniques indicate that 17α,20β-dihydroxy-4-pregnen-3-one (17α,20β-diOHprog) is the natural maturation-inducing steroid in amago salmon, and probably also in other salmonids. A two-cell type model involving two follicular layers for the production of 17α,20β-diOHprog is proposed. In this model, 17α-hydroxyprogesterone produced by the thecal layer in response to gonadotropin is converted to 17α,20β-diOHprog in the granulosa layer where gonadotropin acts to enhance the activity of 20β-hydroxysteroid dehydrogenase. Specific binding of chum salmon gonadotropin has been demonstrated on crude membrane preparations derived from homogenates of purified granulosa cells and thecal layer preparations from post-vitellogenic amago salmon follicles. The action of gonadotropin on granulosa cells is mediated by the adenylate cyclase/cyclic AMP system. Furthermore, our results suggest that gonadotropin causes the *de novo* synthesis of 20β-hydroxysteroid dehydrogenase in the amago salmon granulosa cell through a mechanism dependent on RNA synthesis.

The endocrine control of spermiation in male salmonids is poorly understood. In males of all species studied, plasma levels of 17α,20β-diOHprog were either not detectable or very low during testicular development and rapidly increased at the onset of spermiation. Furthermore, two successive intraperitoneal injections of 17α,20β-diOHprog into non-spermiating amago salmon induced precocious spermiation about one month prior to the normal spermiation period. In contrast, neither 11-ketotestosterone nor testosterone were effective in inducing precocious spermiation. These results are discussed in relation to the possible involvement of 17α,20β-diOHprog in spermiation of male salmonids.

[1] Laboratory of Reproductive Biology, National Institute for Basic Biology, Okazaki, Japan.
[2] Department of Zoology, University of California, Berkeley, CA.
[3] Department of Anatomy, University of Occupational and Environmental Health, School of Medicine, Yahata-Nishiku, Kitakyushu, Japan.

Introduction

Research into the endocrine control of final gamete maturation in salmonids has made great advances over the last few years, particularly in regard to final oocyte maturation. Most of the economically important salmonid species have been investigated in terms of steroid profiles during final oocyte maturation (Fostier et al., 1983; Goetz, 1983), and combined with data from the application of *in vitro* techniques, we are beginning to have a basic understanding of the endocrine actions and interactions occurring during final oocyte maturation. However, in contrast to the wealth of basic information on the female, the endocrine control of final gamete maturation in males has been generally neglected. This paper briefly summarizes the endocrine control of final gamete maturation in salmonids, and describes our recent data on the maturation-inducing steroid of amago salmon (*Oncorhynchus rhodurus*) regarding (1) its identification and characterization, and (2) control of its synthesis by gonadotropin.

Endocrine Control of Final Gamete Maturation in Salmonids

Oocyte maturation: Fully grown oocytes of salmonids are arrested in prophase of the first meiotic division until immediately before final oocyte maturation and ovulation. Hormonal stimulation occurring in cascading series of steps is required to promote the resumption of meiosis which consists of the breakdown of the germinal vesicle (GVBD) (Kanatani and Nagahama, 1980; Goetz, 1983) (Figure 1). The primary hormone involved in triggering oocyte maturation in salmonids, as in other vertebrates, is gonadotropin. Although a number of biochemical studies have been conducted to purify piscine gonadotropins, it is still not clear whether the teleost pituitary contains more than one type of gonadotropins. A glycoprotein-rich gonadotropin with a molecular weight of 25,000-40,000 has been purified in several teleosts (Burzawa-Gerard, 1982; Idler and Ng, 1983). This type of gonadotropin has been reported to stimulate almost all ovarian activities including final oocyte maturation. Measurements of plasma or

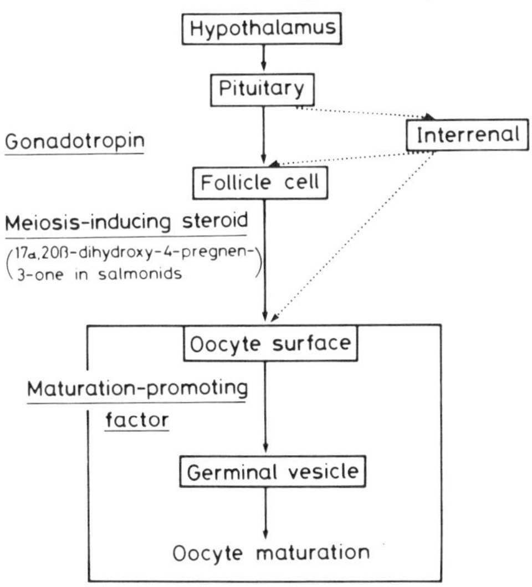

Figure 1. Possible hormonal control of oocyte maturation in teleosts. Hormonal stimulation in cascading series of steps is required to induce oocyte maturation (germinal vesicle breakdown). Maturation-promoting factor has been detected only in goldfish oocytes matured by *in vivo* HCG treatment.

tissue concentrations of teleost gonadotropins reported so far have been made using antisera to this glycoprotein-rich gonadotropin. Recently a second gonadotropin with a low carbohydrate content has also been isolated in certain teleosts. This second gonadotropin termed vitellogenic gonadotropin, has been shown to stimulate the uptake of vitellogenin into the oocyte (Idler and Ng, 1983). In this paper, the term gonadotropin, unless specified otherwise, refers to the glycoprotein-rich gonadotropin.

The increase in plasma gonadotropin levels during oocyte maturation and ovulation has been reported in several species of salmonids (Crim et al., 1973, 1975; Fostier et al., 1978, 1981; Scott et al., 1983; Young et al., 1983a). However, little is known of the initial stimulus causing the rapid elevation of plasma gonadotropin levels which occurs at the time of oocyte maturation. In this connection, it should be noted that in salmonids, plasma estradiol-17β levels are high during the active vitellogenic period, but declined rapidly prior to final oocyte maturation. The drop in plasma estradiol-17β is reflected in the ability of the ovarian follicle to produce this steroid in response to salmon gonadotropin (Kagawa et al., 1983). The loss of ability of the follicle to produce estradiol-17β seems largely due to the loss of aromatase activity in the granulosa cell (Young et al., 1983b). Whether the drop in plasma estradiol-17β levels is related to the increase of gonadotropin levels in the plasma remains to be clarified in salmonids.

Since salmonid oocytes denuded of their follicular envelopes are incapable of responding to gonadotropin, it is believed that gonadotropin does not act directly on the oocyte but acts on their surrounding ovarian follicle to produce a secondary steroidal effector, maturation-inducing steroid (Jalabert, 1976; Kanatani and Nagahama, 1980; Young et al., 1982; Goetz, 1983). In salmonids, corticosteroids and pregnene derivatives such as progesterone, 17α-hydroxyprogesterone, 20β-dihydroprogesterone and 17α, 20β-dihydroxy-4-pregnen-3-one (hereafter abbreviated as 17α,20β-diOHprog) have been shown to be effective in inducing meiotic maturation *in vitro*; among them, 17α,20β-diOHprog has been reported to be the most effective steroid (Fostier et al., 1973; Jalabert, 1976; Duffey and Goetz, 1980; Nagahama et al., 1980, 1983; Sower and Schreck, 1982a). 17α,20β-DiOHprog was first identified in the plasma of spawning sockeye salmon (*Oncorhynchus nerka*) (Idler et al., 1960), and Atlantic salmon (*Salmo salar*) (Schmidt and Idler, 1962). This steroid has been shown to be synthesized by the follicle in response to gonadotropin concomitant with the induction of GVBD (Fostier et al., 1981; Young et al., 1982, 1983a) and elevated concentrations of this steroid are found in the plasma of females undergoing final oocyte maturation (Campbell et al., 1980; Fostier et al., 1981; Scott and Baynes, 1982; Scott et al., 1983; Wright and Hunt, 1982; Young et al., 1983a; Ueda et al., 1984a). 17α,20β-DiOHprog is thus considered to be the natural mediator of gonadotropin-induced oocyte maturation in salmonids.

It is also possible that other steroid hormones are involved in the process of oocyte maturation. High concentrations of plasma testosterone have been found in several salmonids during the spawning period (Scott et al., 1980, 1983; Campbell et al., 1980; Kagawa et al., 1983; Sower and Schreck, 1982b). Furthermore, recent *in vitro* studies with amago salmon have shown that fully grown follicles produce large amounts of testosterone in response to salmon gonadotropin. Although in teleosts, testosterone is effective in inducing *in vitro* final oocyte maturation only at very high concentrations, it has been reported to enhance the effectiveness of gonadotropin on the induction of final oocyte maturation in rainbow trout (*Salmo gairdneri*) (Jalabert, 1976) and amago salmon (Young et al., 1982). Corticosteroids have also been demonstrated to enhance the effect of gonadotropin or steroids on the induction of oocyte maturation in salmonids (Jalabert, 1975; Young et al., 1982). Thus, the exact role of these steroid hormones, particularly the possible involvement of testosterone, in the control of final oocyte maturation in salmonids needs to be investigated.

The involvement of interrenal tissue on final oocyte maturation in salmonids has not been investigated in detail. Furthermore, information is not available in salmonids concerning biochemical events occurring within the oocyte during GVBD in response to the maturation-inducing steroid.

Spermiation: In male salmonids, a marked increase in the relative amount of mature spermatozoa to total germ cells occurs during the later stages of spermatogenesis. Nonetheless, empirical observations show that spermiation does not occur until a majority of cells have completed spermatogenesis. Morphological events occurring in the testis during spermiation have not yet been studied in detail. Spermiation has been considered to be under the hormonal control, and injection of pituitary extracts or gonadotropins stimulates spermiation in several teleosts (Clemens and Grant, 1965; Billard et al., 1982). It is generally assumed that exogenous gonadotropin does not act directly to induce spermiation, but works in concert with testicular somatic elements to stimulate the production of steroidal mediator(s). Although exogenous application of certain steroids, particularly androgens, can induce precocious spermiation in goldfish (*Carassius auratus*) (Yamazaki and Donaldson, 1968; Billard et al., 1982), the role of steroids in inducing spermiation in salmonids is not clear.

Recently, increases in plasma 17α,20β-diOHprog have been demonstrated in spermiating rainbow trout (Scott and Baynes, 1982), amago salmon (Ueda et al., 1983) and chum salmon (*Oncorhynchus keta*) (Ueda et al., 1984a). Furthermore, we have recently found that a single injection of chum salmon gonadotropin (SGA, Syndel Lab. Ltd., Canada) or two successive intraperitoneal injections of 17α,20β-diOHprog are effective in inducing precocious spermiation in amago salmon about one month prior to the normal spermiation period; in these same experiments SGA dramatically increased serum levels of 17α,20β-diOHprog. In contrast, neither testosterone nor 11-ketotestosterone were effective in inducing precocious spermiation (Ueda et al., 1984c). Taken together, these results suggest the possible involvement of 17α,20β-diOHprog in gonadotropin-induced spermiation in salmonids.

Incubation of testicular fragments from spermiating rainbow trout and amago salmon with chum salmon gonadotropin (SGA) or 17α-hydroxyprogesterone resulted in a highly significant increase in 17α,20β-diOHprog levels in the incubation medium (Ueda et al., 1983). This finding indicates that the testes are the major source of 17α,20β-diOHprog. However, the cellular site of synthesis of 17α,20β-diOHprog in the testes has not yet been investigated. Our recent studies have shown that incubation of isolated sperm from spermiating amago salmon produces large amounts of 17α,20β-diOHprog in response to 17α-hydroxyprogesterone (Ueda et al., 1984b).

Furthermore, after incubation of sperm preparations with [^{14}C] 17α-hydroxyprogesterone for 18 hours at 15°C, 17α,20β-diOHprog was identified as the only metabolite (Ueda, unpublished). Considered together, these findings indicate that sperm possess 20β-hydroxysteroid dehydrogenase, the enzyme controlling the conversion from 17α-hydroxyprogesterone to 17α,20β-diOHprog. It is a subject for further study to define the cellular localization and regulatory mechanisms of 20β-hydroxysteroid dehydrogenase in sperm, and to investigate the first appearance of this enzyme during germ cell development. Moreover, whether 20β-hydroxysteroid dehydrogenase activity is limited only to sperm or whether it is also present in other testicular tissues is open to question.

Maturation-inducing Steroid of Amago Salmon

Isolation and characterization: It has been shown that folliculated fully grown oocytes of amago salmon undergo GVBD *in vitro* when they are incubated with chinook salmon gonadotropin (SG-G100) (Nagahama et al., 1980). Further investigations using cyanoketone, a specific inhibitor of 3β-hydroxy-Δ^5-steroid dehydrogenase, have demonstrated that the action of gonadotropin is dependent on the synthesis of a second steroidal mediator of oocyte maturation (Young et al., 1982). The maturation-inducing steroid of amago salmon was purified and characterized from media in which immature but fully grown folliculated oocytes of amago salmon had been incubated for 18-24 hours with chum salmon gonadotropin (SGA).

Ether extracts of the media from these incubates showed high maturation-inducing (MI) activity; MI activity was assessed by an *in vitro* GVBD assay using fully grown prophase-arrested oocytes of amago salmon. Yolk and oil droplets were removed from the ether extract by partition with equal volumes of 50% methanol and n-hexane. MI activity was found only in the 50% methanol phase. The 50% methanol phase was then fractionated (20 separate fractions) by reverse phase high performance liquid chromatography. MI activity was found only in fraction 10 which had a retention time coinciding exactly with 17α,20β-DiOHprog (Figure 2). The purity and final characterization of the residue of fraction 10 were further confirmed by thin layer chromatography and mass spectroscopy with authentic 17α,20β-DiOHprog standard (Figure 3).

Experiments examining the relative effectiveness of a range of pregnene derivatives in inducing GVBD in amago salmon have shown that 17α,20β-diOHprog is the most potent inducer of oocyte maturation (Nagahama et al., 1983). 17α,20β-DiOHprog is one of two major metabilites of [^{14}C]-17α-hydroxyprogesterone produced by cell-free ovarian homogenates of amago salmon (Suzuki et al., 1981). The other major metabolite, 17α-hydroxy-5β-pregnane-3,20-dione is a relatively ineffective inducer of GVBD in amago salmon oocytes *in vitro* (Nagahama et al., 1983). A radioimmunoassay for 17α,20β-diOHprog recently developed in our laboratory was used to investigate the relationship between the occurrence of this steroid and gonadotropin in the plasma of female amago salmon during sexual maturation. Both gonadotropin and 17α,20β-diOHprog levels were low in vitellogenic females and in those with fully grown immature oocytes and were strikingly elevated in mature and ovulated females (Young et al., 1983a). *In vitro* production of 17α,20β-diOHprog by ovarian follicles at different stages of development has clearly shown that the capacity of the follicles to respond to gonadotropin by synthesizing and secreting this steroid is acquired immediately prior to the natural maturation period (Young et al., 1983a). Furthermore, 17α,20β-diOHprog seems to be an end product, since no further metabolites were produced by cell-free ovarian homogenates of amago salmon when this steroid was used as substrate (Tamaoki et al., 1984). The preceding findings lead to the conclusion that 17α,20β-diOHprog is the natural maturation-inducing steroid in amago salmon. Further investigations from our laboratory using similar techniques suggest that 17α,20β-diOHprog is the natural maturation-inducing steroid in the other three species of salmonids, chum salmon, masu salmon (*Oncorhynchus masou*), and rainbow trout.

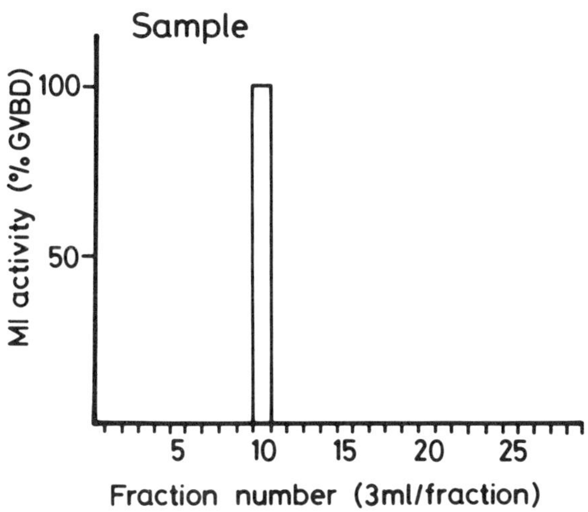

Figure 2. Maturation-inducing (MI) activity of various fractions of the 50% methanol phase separated by high performance liquid chromatography. MI activity is found only in fraction 10, where 100% germinal vesicle breakdown (GVBD) is recorded with undiluted sample.

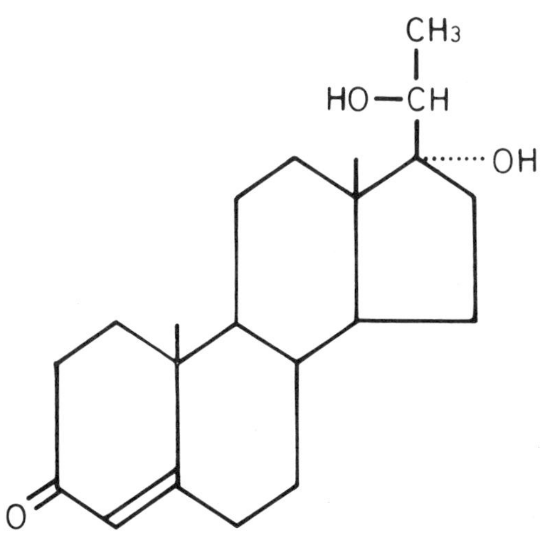

Figure 3. Maturation-inducing steroid of the amago salmon. 17α,20β-dihydroxy-4-pregnen-3-one.

Two-cell type model for the production of 17α,20β-diOHprog: The identification of the maturation-inducing steroid in amago salmon and rainbow trout as 17α,20β-diOHprog permitted a study of the role of the follicle layer in the production of this steroid. The ovarian follicle layer of teleosts, as in that of other vertebrates, consists of two major layers: an inner granulosa layer and an outer thecal layer (Nagahama et al., 1982; Nagahama, 1983). Taking advantage of characteristics of the salmonid ovarian follicle, we have been attempting to elucidate the role of the thecal and granulosa cells in the production of maturation-inducing steroid in amago salmon and rainbow trout. Using recently developed *in vitro* techniques, similar to those used for the studies on follicular estradiol-17β production (Kagawa et al., 1982, 1984; Nagahama et al., 1984), three follicular preparations (thecal layer, granulosa layer, and co-culture of thecal and granulosa layers) were obtained from fully grown oocytes of rainbow trout and amago salmon and were incubated with or without partially purified chinook salmon gonadotropin (SG-G100). 17α-Hydroxyprogesterone and 17α,20β-diOHprog levels in the media were measured by specific radioimmunoassay. *In vitro* production of 17α,20β-diOHprog by intact follicles and co-culture preparations was remarkably enhanced by SG-G100, but neither isolated thecal layers nor isolated granulosa layers were capable of producing substantial amounts of 17α,20β-diOHprog in response to this gonadotropin. These *in vitro* data indicate that the interaction of both thecal and granulosa layers is necessary for the production of 17α,20β-diOHprog in response to SG-G100.

The concentration of 17α-hydroxyprogesterone of media from the same experiment was determined. Isolated thecal layers produced large amounts of 17α-hydroxyprogesterone in response to SG-G100, but no stimulation was recorded in incubates with isolated granulosa layers. In contrast, levels of 17α-hydroxyprogesterone in media from intact follicle and co-culture incubations peaked at 12 hours and rapidly decreased, concomitant with a rapid rise in 17α,20β-diOHprog levels. Furthermore, when isolated granulosa layers were incubated with exogenous 17α-hydroxyprogesterone, they produced 17α,20β-diOHprog, thereby indicating the presence of 20β-hydroxysteroid dehydrogenase in granulosa layers. It has also been shown that chinook salmon gonadotropin (SG-G100), although alone unable to stimulate 17α,20β-diOHprog production, strikingly enhanced 17α,20β-diOHprog production when 17α-hydroxyprogesterone was present in the incubation medium. This observation can be interpreted as a direct action of SG-G100 on the enhancement of 20β-hydroxysteroid dehydrogenase activity. Considering these data together, a two-cell type model for the production of the maturation-inducing steroid by the salmonid ovarian follicle has been proposed, which is, to our knowledge the first time in any vertebrate (Nagahama, 1983; Young et al., 1984). In this model, under the influence of gonadotropin, the thecal layer synthesizes precursors, probably 17α-hydroxyprogesterone, which are transferred to the granulosa layer and coverted to 17α,20β-diOHprog (Figure 4). It is also possible that, in addition to gonadotropin, steroids can modulate their own synthesis or the synthesis of other steroids in the ovarian follicle, possibly through a direct interaction with enzymes involved in steroidogenesis.

Gonadotropin receptors in the ovarian follicle: In the two-cell type model for the production of 17α,20β-diOHprog described above, gonadotropin has at least

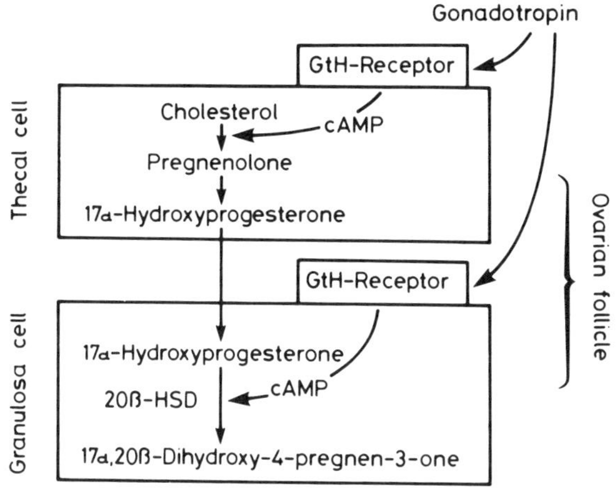

Figure 4. Two-cell type model for the production of follicular 17α,20β-dihydroxy-4-pregnen-3-one in the amago salmon (see text for details). 20β-HSD, 20β-hydroxysteroid dehydrogenase.

two sites of action: on the thecal layer to stimulate the production of 17α-hydroxyprogesterone; and on the granulosa layer to enhance the activity of 20β-hydroxysteroid dehydrogenase. In collaboration with C. Salmon (Prof. Y.A. Fontaine's laboratory at the Muséum national d'Histoire naturelle, France), we have been investigating gonadotropin binding to both thecal and granulosa layers of amago salmon. Purified carp gonadotropin (purified by E. Burzawa-Gérard) and chum salmon gonadotropin (purified by H. Ueda) were iodinated with the aid of Iodogen (1,3,4,6-tetrachloro-3α,6α-diphenylglycoluril). Crude membrane preparations derived from the 23,000 g pellet of homogenates of purified granulosa cells and thecal layer preparations were obtained from post-vitellogenic amago salmon follicles. [^{125}I]-Salmon gonadotropin showed a relatively high specific binding (5-7% of total radioactivity) with relatively low non-specific binding (1-2%) to granulosa cell preparations (Salmon et al., 1984). However, non-specific binding was much higher with thecal layer preparations and specific binding was lower; this may be related to the heterogeneous nature of the thecal tissue. The binding studies with [^{125}I]-carp gonadotropin generally showed relatively high non-specific and a low specific binding to ovarian follicular preparations of amago salmon. No specific binding was found with a variety of non-gonadal tissues including the brain, kidney, muscle, and liver. Further investigations using these techniques are required to elucidate the sites and characteristics of gonadotropin binding and for understanding the endocrine regulation of the fish gonadotropin receptors.

Adenylate cyclase/cyclic AMP system: The functional role of the andenylate cyclase/cyclic AMP system in the gonadotropin-induced enhancement of 20β-hydroxysteroid dehydrogenase activity in the amago salmon granulosa layer has been investigated. When granulosa layers were incubated with forskolin, an adenylate cyclase activator, in the absence of 17α-hydroxyprogesterone, no 17α,20β-diOHprog was produced. In contrast, in the presence of 17α-hydroxyprogesterone, the production of 17α,20β-diOHprog was strikingly stimulated by forskolin. Similarly, dibutyryl cyclic AMP and two phosphodiesterase inhibitors, theophylline and 3-isobutyl-l-methylxanthine (IBMX), strikingly enhanced the production of 17α, 20β-diOHprog when 17α-hydroxyprogesterone was present in the incubation medium. Furthermore, when isolated granulosa layers were incubated for 1 hour with or without chum salmon gonadotropin (SGA) in the presence of IBMX, about three times more cyclic AMP was produced in incubates with gonadotropin, and there was a good correlation between the number of granulosa layers per sample and cyclic AMP concentration. Taken together, these results suggest that in the amago salmon granulosa layer, the adenylate cyclase/cyclic AMP system acts as an intracellular mediator in the activation of 20β-hydroxysteroid dehydrogenase by gonadotropin.

Effects of cycloheximide and actinomycin D: To elucidate the molecular mechanisms involved in the gonadotropin-induced 20β-hydroxysteroid dehydrogenase activation, we have examined the effects of cycloheximide and actinomycin D on the production of 17α,20β-diOHprog by gonadotropin in the amago salmon granulosa layer. When varying concentrations (10 - 0.001 μg/ml) of either cycloheximide or actinomycin D were added to the incubation medium containing 17α-hydroxyprogesterone, stimulatory effects of chum salmon gonadotropin (SGA) on the production of 17α,20β-diOHprog were significantly inhibited. Similarly, the dibutyryl cyclic AMP-induced 20β-hydroxysteroid dehydrogenase activation was also inhibited by both cycloheximide and actinomycin D. Thus, these results suggest that gonadotropin causes the *de novo* synthesis of 20β-hydroxysteroid dehydrogenase in the amago salmon granulosa layer through a mechanism dependent on RNA synthesis.

In summary the proposed mechanisms involved in the activation of 20β-hydroxysteroid dehydrogenase by gonadotropin in the amago salmon granulosa cell are as follows. Gonadotropin first binds to the receptors on the granulosa cell membrane and activates the adenylate cyclase/cyclic AMP system. Since cycloheximide and actinomycin D block both the gonadotropin and cyclic AMP-induced 20β-hydroxysteroid dehydrogenase enhancement, we suggest that gonadotropin promotes *de novo* synthesis of 20β-hydroxysteroid dehydrogenase by a mechanism involving RNA synthesis. We are attempting to purify 20β-hydroxysteroid dehydrogenase to allow us to investigate further the molecular events involved in gonadotropin enhancement of this enzyme, which is a key step in the oocyte maturation process. Clearly several parts of this proposed model require further investigation. While our model relates specifically to the amago salmon, it should provide a basis for our understanding of the molecular action of gonadotropin in other vertebrates.

References

Billard, R., Fostier, A., Weil, C. and Breton, B., 1982. Endocrine control of spermatogenesis in teleost fish. Can. J. Fish. Aquat. Sci., 39:65-79.

Burzawa-Gerard, E., 1982. Chemical data on pituitary gonadotropins and their implication to evolution. Can. J. Fish. Aquat. Sci., 39:80-91.

Campbell, C.M., Fostier, A., Jalabert, B. and Truscott, B., 1980. Identification and quantification of steroids in the serum of rainbow trout during spermiation and oocyte maturation. J. Endocrinol., 85:371-378.

Clemens, H.P. and Grant, F.B., 1965. The seminal thinning response of carp (*Cyprinus carpio*) and rainbow trout (*Salmo gairdneri*) after injections of pituitary extracts. Copeia, 2:174-177.

Crim, L.W., Meyer, R.K. and Donaldson, E.M., 1973. Radioimmunoassay estimates of plasma gonadotropin levels in the spawning pink salmon. Gen. Comp. Endocrinol., 21:69-76.

Crim, L.W., Watt, E.G. and Evans, D.M., 1975. The plasma gonadotropin profile during sexual maturation in a variety of salmonid fishes. Gen. Comp. Endocrinol., 27:62-70.

Duffey, R.J. and Goetz, F.W., 1980. The *in vitro* effects of 17α-hydroxy-20β-dihydroprogesterone on germinal vesicle breakdown in brook trout (*Salvelinus fontinalis*) oocytes. Gen. Comp. Endocrinol., 41:563-565.

Fostier, A., Breton, B., Jalabert, B. and Marcuzzi, O., 1981. Évolution des niveaux plasmatiques de la gonadotropine glycoprotèique et de la 17α-hydroxy-20β-dihydroprogèsterone au cours de la maturation et de l'ovulation chez la Truite arc-en-ciel, *Salmo gairdneri*. C. R. Acad. Sci. Paris, Sèr. III 293:817-820.

Fostier, A., Jalabert, B. and Terqui, M., 1973. Action predominante d'un dèrivè hydroxylè de la progestèrone sur la maturation *in vitro* des ovocytes de la Truite arc-en-ciel, *Salmo gairdnerii*. C. R. Acad. Sci. Paris, D 277:421-424.

Fostier, A., Jalabert, B., Billard, B., Breton, B. and Zohar, Y., 1983. The gonadal steroids. In: W.S. Hoar, D.J. Randall and E.M. Donaldson (editors), Fish Physiology, Vol. IXA, Reproduction. Academic Press, New York, pp. 277-372.

Fostier, A., Weil, C., Terqui, M., Breton, B. and Jalabert, B., 1978. Plasma estradiol-17β and gonadotropin during ovulation in rainbow trout (*Salmo gairdneri* R.). Ann. Biol. anim. Bioch. Biophys., 18:929-936.

Goetz, F.W., 1983. Hormonal control of oocyte final maturation and ovulation in fishes. In: W.S. Hoar, D.J. Randall and E.M. Donaldson (editors), Fish Physiology, Vol. IXB, Reproduction. Academic Press, New York, pp. 117-170.

Idler, D.R., Fagerlund, U.H.M. and Ronald, A.P., 1960. Isolation of pregn-4-ene-17α,20β-diol-3-one from plasma of Pacific salmon (*Oncorhynchus nerka*). Biochem. Biophys. Res. Commun., 2:133-137.

Idler, D.R. and Ng, T.B., 1983. Teleost gonadotropins: Isolation, biochemistry, and function. In: W.S. Hoar, D.J. Randall and E.M. Donaldson (editors), Fish Physiology, Vol. IXA, Reproduction. Academic Press, New York, pp. 187-221.

Jalabert, B., 1975. Modulation par différent stèröides non maturants de l'efficacité de la 17α-hydroxy-20β-dihydroprogestérone ou d'un extrait gonadotrope sur la maturation intrafolliculare in vitro des ovocytes de la Truite arc-en-ciel *Salmo gairdneri*. C. R. Acad. Sci. Paris, D 281:811-814.

Jalabert, B., 1976. *In vitro* oocyte maturation and ovulation in rainbow trout (*Salmo gairdneri*), northern pike (*Esox lucius*), and goldfish (*Carassius auratus*). J. Fish. Res. Board Can., 33:974-988.

Kagawa, H., Young, G., Adachi, S. and Nagahama, Y., 1982. Estradiol-17β production in amago salmon (*Oncorhynchus rhodurus*) ovarian follicles: Role of the thecal and granulosa cells. Gen. Comp. Endocrinol., 47:440-448.

Kagawa, H., Young, G. and Nagahama, Y., 1983. Relationship between seasonal plasma estradiol-17β and testosterone levels and *in vitro* production by ovarian follicles of amago salmon (*Oncorhynchus rhodurus*). Biol. Reprod., 29:301-309.

Kagawa, H., Young, G., Adachi, S. and Nagahama, Y., 1984. Estrogen synthesis in the teleost ovarian follicle: the two-cell type model in salmonids. This issue, Sea Grant Proceedings.

Kanatani, H. and Nagahama, Y., 1980. Mediators of oocyte maturation. Biomed. Res., 1:273-291.

Nagahama, Y., 1983. The functional morphology of teleost gonads. In: W.S. Hoar, D.J. Randall and E.M. Donaldson (editors), Fish Physiology, Vol. IXA, Reproduction. Academic Press, New York, pp. 223-275.

Nagahama, Y., Hirose, K., Young, G., Adachi, S., Suzuki, K. and Tamaoki, B., 1983. Relative *in vitro* effectiveness of 17α,20β-dihydroxy-4-pregnen-3-one and other pregnene derivatives on germinal vesicle breakdown in oocytes of ayu (*Plecoglossus altivelis*), amago salmon (*Oncorhynchus rhodurus*), rainbow trout (*Salmo gairdneri*), and goldfish (*Carassius auratus*). Gen. Comp. Endocrinol., 51:15-23.

Nagahama, Y., Kagawa, H. and Tashiro, F., 1980. The *in vitro* effects of various gonadotropins and steroid hormones on oocyte maturation in amago salmon *Oncorhynchus rhodurus* and rainbow trout *Salmo gairdneri*. Bull. Japan. Soc. Sci. Fish., 39:765-769.

Nagahama, Y., Kagawa, H. and Young, G., 1982. Cellular sources of sex steroids in teleost gonads. Can. J. Fish. Aquat. Sci., 39:56-64.

Nagahama, Y., Young, G. and Kagawa, H., 1984. Steroidogenesis in the amago salmon (*Oncorhynchus rhodurus*) ovarian follicle: a two-cell type model. Proc. 9th Int. Symp. Comp. Endocrinol., (in press).

Salmon, C., Kagawa, H., Nagahama, Y. and Fontaine, Y., 1984. C. R. Acad. Sci. Paris, Sèr. III, 298:331-334.

Schmidt, P.J. and Idler, D.R., 1962. Steroid hormones in the plasma of salmon at various states of maturation. Gen. Comp. Endocrinol., 2:204-214.

Scott, A.P. and Baynes, S.M., 1982. Plasma levels of sex steroids in relation to ovulation and spermiation in rainbow trout (*Salmo gairdneri*). In: C.J.J. Richter and H.J.Th. Goos (editors), Reproductive Physiology, Center for Agricultural Publishing and Documentation (Pudoc), Wageningen, pp. 103-106.

Scott, A.P., Bye, V.J. and Baynes, S.M., 1980. Seasonal variations in sex steroids of female rainbow trout (*Salmo gairdneri* Richardson). J. Fish Biol., 17:587-592.

Scott, A.P., Sheldrick, E.L. and Flint, A.P.F., 1983. Measurement of 17α,20β-dihydroxy-4-pregnen-3-one in plasma of trout (*Salmo gairdneri* Richardson): Seasonal changes in response to salmon pituitary extract. Gen. Comp. Endocrinol., 46:444-451.

Sower, S.A. and Schreck, C.B., 1982a. *In vitro* induction of final oocyte maturation of oocytes in coho salmon. Trans. Amer. Fish. Soc., 111:399-403.

Sower, S.A. and Schreck, C.B., 1982b. Steroid and thyroid hormones during sexual maturation of coho salmon (*Oncorhynchus kisutch*) in seawater or fresh water. Gen. Comp. Endocrinol. 47: 42-53.

Suzuki, K., Nagahama, Y. and Tamaoki, B., 1981. *In vitro* synthesis of an inducer for germinal vesicle breakdown of fish oocytes, 17α,20β-dihydroxy-4-pregnen-3-one by ovarian tissue preparation of amago salmon (*Oncorhynchus rhodurus*). Gen. Comp. Endocrinol., 45:533-535.

Tamaoki, B., Suzuki, K., Hirose, K. and Nagahama, Y., 1984. Biosynthesis of 17α,20β-dihydroxy-4-pregnen-3-one, an inducer for germinal vesicle breakdown of fish oocytes by teleostean ovaries. Proc. 9th Int. Symp. Comp. Endocrinol., (in press).

Ueda, H., Hiroi, O., Hara, A., Yamauchi, K. and Nagahama, Y., 1984a. Changes in serum concentrations of steroid hormones, thyroxine, and vitellogenin during spawning migration of chum salmon, *Oncorhynchus keta*. Gen. Comp. Endocrinol., 53:203-211.

Ueda, H., Kambegawa, A. and Nagahama, Y., 1984b. *In vitro* 11-ketotestosterone and 17α,20β-dihydroxy-4-pregnen-3-one production by testicular fragments and isolated sperm of rainbow trout, *Salmo gairdneri*. J. Exp. Zool., 231:435-439.

Ueda, H., Kambegawa, A. and Nagahama, Y., 1984c. Involvement of gonadotrophin and steroid hormones in spermiation in the amago salmon *(Oncorhynchus rhodurus)* and goldfish *Carassius auratus*. Gen. Comp. Endocrinol. (in press).

Ueda, H., Young, G., Crim, L.W., Kambegawa, A. and Nagahama, Y., 1983. 17α,20β-Dihydroxy-4-pregnen-3-one: Plasma levels during sexual maturation and *in vitro* production by the testes of amago salmon (*Oncorhynchus rhodurus*) and rainbow trout (*Salmo gairdneri*). Gen. Comp. Endocrinol., 51:106-112.

Wright, R.S. and Hunt, S.M.V., 1982. A radioimmunoassay for 17α,20β-dihydroxy-4-pregnen-3-one: Its use in measuring changes in serum levels at ovulation in Atlantic salmon (*Salmo salar*), coho salmon (*Oncorhynchus kisutch*), and rainbow trout (*Salmo gairdneri*). Gen. Comp. Endocrinol., 47:475-482.

Yamazaki, F. and Donaldson, E.M., 1968. Involvement of gonadotropin and steroid hormones in the spermiation of the goldfish (*Carassius auratus*). Gen. Comp. Endocrinol., 12:491-497.

Young, G., Kagawa, H. and Nagahama, Y., 1982. Oocyte maturation in the amago salmon (*Oncorhynchus rhodurus*): *In vitro* effects of salmon gonadotropin, steroids, and cyanoketone (an inhibitor of 3β-hydroxy-Δ^5-steroid dehydrogenase). J. Exp. Zool., 224:265-275.

Young, G., Crim, L.W., Kagawa, H., Kambegawa, A. and Nagahama, Y., 1983a. Plasma 17α,20β-dihydroxy-4-pregnen-3-one levels during sexual maturation of amago salmon (*Oncorhynchus rhodurus*): Correlation with plasma gonadotropin and *in vitro* production by ovarian follicles. Gen. Comp. Endocrinol., 51:96-105.

Young, G., Kagawa, H. and Nagahama, Y., 1983b. Evidence for a decrease in aromatase activity in the ovarian granulosa cells of amago salmon (*Oncorhynchus rhodurus*) associated with final oocyte maturation. Biol. Reprod., 29:310-315.

Young, G., Kagawa, H. and Nagahama, Y., 1984. Role of the thecal and granulosa cells in the production of maturation-inducing steroid (MIS) by ovarian follicles of salmonid fishes. Gen. Comp. Endocrinol., 53:455.

Acknowledgments

The authors wish to thank Mr. F. Tashiro, Director of the Gifu Prefectural Fisheries Experimental Station for providing amago salmon. The studies from our laboratory described in this article have been aided in part by Grant-in-Aid for special Project Research (Project No. 58119008) and Scientific Research from the Ministry of Education, Science and Culture, Japan.

ESTROGEN SYNTHESIS IN THE TELEOST OVARIAN FOLLICLE

The Two-cell Type Model in Salmonoids

Hirohiko Kagawa,[1] *Graham Young,*[2] *Shinji Adachi,*[3] *and Yoshitaka Nagahama*[3]

Abstract: Studics conducted to assess the role of the ovarian thecal and granulosa layers of several species of salmonoids, particularly of amago salmon, *Oncorhynchus rhodurus*, during estradiol-17β (E_2) production are discussed. Four different follicular preparations from vitellogenic follicles of amago salmon (intact follicles, thecal layers, granulosa layers, and thecal layer-granulosa layer co-cultures) were incubated in the presence or absence of partially purified salmon gonadotropin (SG-G100). E_2 and testosterone (T) levels in the media were measured by specific radioimmunoassay. SG-G100 stimulated E_2 production by intact follicles and co-culture preparations, but not by the isolated thecal or granulosa layers, indicating that both layers are necessary for gonadotropin-stimulated E_2 production. In contrast, SG-G100 greatly stimulated T production by thecal layers but only slightly the production of other follicular preparations. Experiments used cyanoketone, a specific inhibitor of 3β-hydroxy-Δ^5-steroid dehydrogenase, suggest that E_2 precursors secreted by the thecal layer are Δ^4 androgens. These Δ^4 androgens have been identified as T and androstenedion (A) by using reversed phase high performance liquid chromatography, thin layer chromatography, and mass spectoscopy. These results indicate a two-cell type model for the production of follicular E_2 in amago salmon, the thecal layer contributing to E_2 production by synthesizing Δ^4 androgens (A and T) which are aromatized in the granulosa layer to E_2. A two-cell type model also seems applicable to the production of follicular estrogen in other species of salmonoids (chum and masu salmon, and rainbow trout).

Introduction

Abundant evidence exists for steroid secretion by the ovary of fishes (Ozon, 1972). As to streroidogenesis, it has been demonstrated that the ovary of teleosts is capable of synthesizing estrogens, androgens, progestogens, and also, in a few restricted species, corticosteroids by *in vitro* incubation with radiolabeled steroid precursors. Studies using histochemical (Guraya, 1976) and ultrastructural (Nagahama et al., 1982) methods suggest that the granulosa and/or thecal cells are the main sites of steroid production, depending upon the species

[1] Department of Anatomy, University of Occupational Health, School of Medicine, Yahata-Nishiku, Kitakyushi, Japan
[2] Department of Zoology, and Cancer Research Laboratory, University of California, Berkeley, CA
[3] Laboratory of Reproductive Biology, National Institute for Basic Biology, Okazaki, Japan

examined. However, direct evidence on the mechanism of steroid production in the teleost ovarian follicle has until recently been lacking. In this paper, we will review information about the mechanisms underlying estrogen synthesis in the salmonoid ovarian follicle, particularly of the amago salmon (*Oncorhynchus rhodurus*), obtained using *in vitro* incubation techniques and steroid radioimmunoassay, and discuss some aspects of the control of steroidogenesis in the teleost ovarian follicle.

In Vitro Steroid Production in Response to Gonadotropin

In the teleost fishes, estradiol-17β (E_2) is one of the most important steroids involved in stimulating the synthesis and secretion of hepatic vitellogenic proteins; the increase of plasma E_2 levels during vitellogenesis is well documented (Lance and Callard, 1978). Although it is generally accepted that in teleosts as in mammals pituitary gonadotropin(s) acts directly upon the follicle to stimulate E_2 production, the only direct evidence in salmonids comes from our *in vitro* studies on the amago salmon. Isolated oocytes with follicle layers (intact follicles), which are obtained from the fish during early or mid vitellogenesis (the period of June to August), were incubated with partially purified chinook salmon gonadotropin (SG-G100) for 18 hours at 15°C. Following incubation, E_2 levels were measured by radioimmunoassay. The accumulation of E_2 in both tissues and media was stimulated by SG-G100 in a dose-dependent fashion (Kagawa et al., 1982a). This result provides direct evidence that gonadotropin can directly stimulate E_2 production by the ovarian follicles of salmon.

Role of Thecal and Granulosa Layers

The teleost ovarian follicle is composed of an inner granulosa layer separated by a basement membrane from an outer thecal layer. Separation of the thecal and granulosa layers facilitates the assessment of their roles in steroid production using *in vitro* incubation methods. The follicle layers of amago salmon are easily separated into two layers by a simple dissection procedure. Using these preparations, we examined the effects of SG-G100 on E_2 production. SG-G100 stimulated E_2 production by intact follicles and thecal and granulosa layer co-culture preparations, but not by the isolated thecal or granulosa layers (Figure 1). These results indicated that both layers are necessary for gonadotropin-stimu-

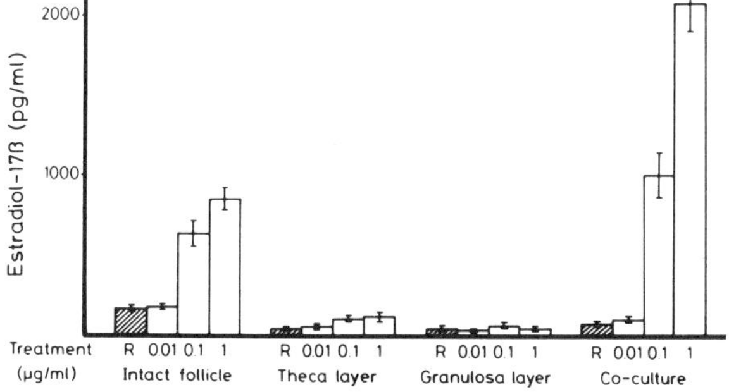

Figure 1. Effect of chinook salmon gonadotropin on estradiol-17β secretion by amago salmon follicles. Follicles were incubated in Ringer alone (R, shaded bars) or Ringer with various doses of chinook salmon gonadotropin (0.01-1 µg+/ml) (white bars) for 18 hours. The vertical bars represent the mean SEM of the three replicates. (Kagawa et al., 1982b.)

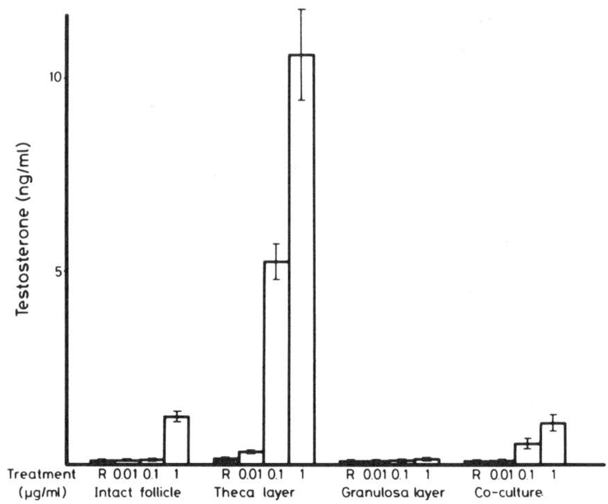

Figure 2. Effects of chinook salmon gonadotropin on testosterone secretion by amago salmon follicles. Follicles were incubated in Ringer alone (R, shaded bars) or Ringer with various doses of chinook salmon gonadotropin (0.01-1 μg/ml) (white bars) for 18 hours. The vertical bars represent the mean SEM of the three replicates. (Kagawa et al., 1982b.)

lated E_2 production. In contrast, SG-G100 greatly stimulated testosterone (T) production by thecal layers but only slightly stimulated T production by the other follicular preparations (Figure 2). Incubation of granulosa layers with exogenous T resulted in elevated E_2 levels, whereas isolated thecal layers incubated with T produced relatively small amounts of E_2 which should be attributed to contamination of thecal layer preprations with granulosa cells (Kagawa et al., 1982b). These results suggested a two-cell type model for the production of follicular estrogens, the thecal layer possibly contributing to E_2 production by synthesizing aromatizable androgens which are transferred to the granulosa layer and aromatized to E_2.

Identification of Aromatizable Androgens

We further investigated the role of the thecal layer and the granulosa later in E_2 production using cyanoketone, a specific inhibitor of 3β-hydroxy-Δ^5-steroid dehydrogenase (3β-HSD) (Young et al., 1982a,b). Thecal layers incubated with SG-G100 secreted large amounts of T but not E_2 into incubation medium. After incubation of granulosa layers in this medium E_2 levels increased. Cyanoketone inhibited SG-G100-induced T production of thecal layers and also inhibited E_2 production by granulosa layers when they are incubated in this medium. However, addition of cyanoketone only to the granulosa layer incubations did not inhibit E_2 production. Furthermore, *in vitro* incubation experiments with various steroids showed that granulosa layers have a limited capacity to metabolize exogenous pregnenolone, 17α-hydroxypregnenolone, progesterone, and 17α-hydroxyprogesterone, but show evidence of strong activity of two enzymes, 17β-hydroxysteroid dehydrogenase and 3β-HSD in addition to aromatase. These experiments suggests first that Δ^5 steroids produced by thecal layers do not make a significant contribution to E_2 production by granulosa layers and second that aromatizable Δ^4 androgens produced by the thecal layers are T and probably androstenedione (A). The presence of 3β-HSD in the granulosa layer remains enigmatic.

Recently, using biochemical methods, we have conclusively identified the aromatizable androgens produced by thecal layers in response to SG-G100. Thecal layers were incubated with SG-G100 and ether extracts of media from these incubates were fractionized (4.5 ml/tube, 27 separated fractions) by reversed phase high performance liquid chromatography. Only two fractions added to the incubation medium were converted to E_2 by granulosa layers. These two fractions had retention times that coincided with authentic A and T respectively. The purity and final characterization of these two fractions were confirmed by thin layer chromatography and mass spectroscopy (Adachi et al., unpublished).

Relationship Between Plasma E_2 and *In Vitro* Production

In amago salmon, E_2 levels in the plasma increase during vitellogenesis and rapidly decline prior to final ocyte maturation (Figure 3). This *in vivo* seasonal pattern is reflected in the ability of the ovarian follicle to produce E_2 in

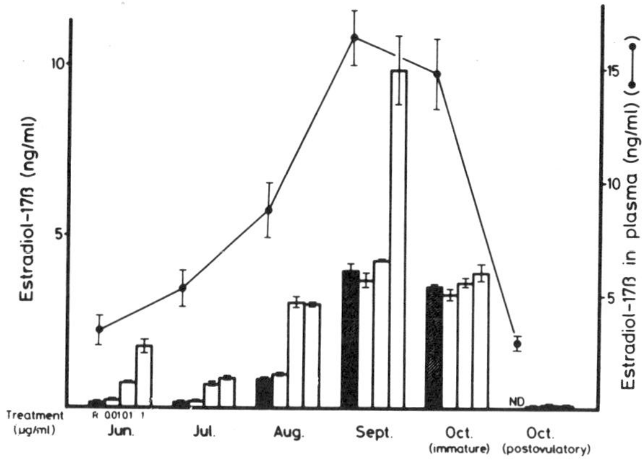

Figure 3. Plasma estradiol-17β levels and its *in vitro* production in response to chinook salmon gonadotropin by ovarian follicles of the amago salmon during sexual maturation. Follicles were incubated in Ringer alone (R, shaded bars) or Ringer with various doses of chinook salmon gonadotropin (0.01-1 µg/ml) (white bars) for 18 hours. The vertical bars represent the mean SEM of the three replicates. (Kagawa et al., 1983.)

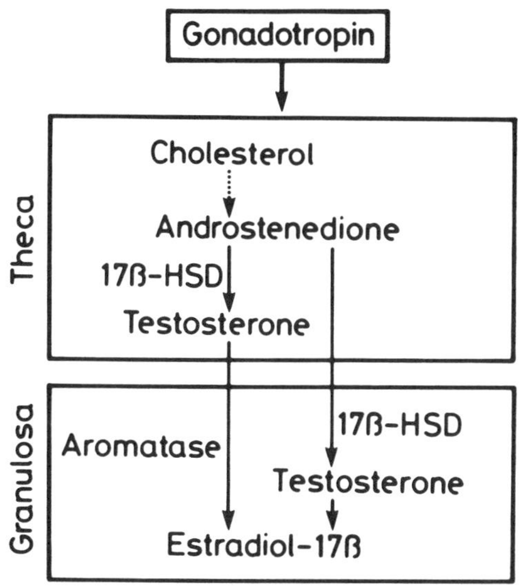

Figure 4. Two-cell model for the synthesis of follicular estradiol-17β in the amago salmon.

response to SG-G100 *in vitro* (Figure 3). Aromatase activity, assessed by the capacity of isolated granulosa layers to produce E_2 from exogenous T, also increases during vitellogenesis and thereafter rapidly declines in the postvitellogenic period (Young et al., 1983). The cause of decreased aromatase activity is not yet established, but it is apparently not due to decreased T, since T production by ovarian follicles remains elevated during the postvitellogenic period (Kagawa et al., 1983; Young et al., 1983).

Two-Cell Type Model of E_2 Production in the Salmonids

We propose a two-cell type model for E_2 production in amago salmon based mainly on the use of *in vitro* incubation techniques (Figure 4). Recently, using the same incubation system and separated follicular preparations, we have found evidence which indicates that this two-cell type model is applicable to the production of follicular estrogen in several other species of salmonids, the chum salmon (*Oncorhynchus keta*), the masu salmon (*O. masou*), the rainbow trout (*Salmo gairdneri*). Thus, the model seems to be generally applicable to salmonid species. This two-cell type model for the production of follicular estrogen is the first report in lower vertebrates. In mammals, the interdependence of thecal and granulosa cells for ovarian estrogen production, first proposed by Falck (1959), is now well-documented (Dorrington and Armstrong, 1979). There appears to be only one other report of a two-cell type model for a non-mammalian vertebrate. Huang and Nalvandov (1979) suggest that the granulosa cells of chickens produce progesterone (or T) which is converted to estrogens by the thecal cells, a finding which is in sharp contrast to salmonids and mammals and of evolutionary interest.

The question of the physiological significance of 3β-HSD in the granulosa cells remain unanswered, since our studies indicate that thecal layers produce only Δ^4 androgens as substrates for E_2. We must therefore emphasize that the present data (indicating that the presence of 3β-HSD in the granulosa cells of amago salmon has only limited significance for E_2 production) is applicable only to the *in vitro* system; its significance during *in vitro* E_2 production remains to be clarified.

It is at present difficult to explain precisely what mechanisms regulate the change in aromatase activity in the granulosa layer. Although aromatase activity increased during vitellogenesis and rapidly declined in the postvitellogenic period, gonadotropin did not affect the metabolism of exogenous T to E_2 by the granulosa layer of amago salmon (Kagawa et al., 1982b, Young et al., 1983). Many of the factors regulating ovarian steroidogenesis proposed in mammals may also apply to teleost ovarian tissue. LH appears to stimulate androgen production by thecal layers (Hamberger et al., 1978) and FSH seems to be involved in the conversion of androgens to E_2 (Dorrington and Armstrong, 1979). FSH in combination with E_2 stimulated a rapid increase in the number of fish receptors per granulosa cell and, later, a distinct increase in LH receptors in granulosa cells in developing follicles (Richards et al., 1976). Peluso et al. (1979) suggest that androgens maintain LH binding sites and prevent the degeneration of the preovulatory follicle. Thus, a reasonable postulation is that changes in receptor content in the follicle cells might determine the response of teleost follicles to specific hormones. We are currently studying the regulation of steroid and gonadotropin receptors in an effort to further clarify the mechanisms controlling steroidogenesis in the teleost ovarian follicle.

References

Dorrington, J.H. and Armstrong, D.T., 1979. Effects of FSH on gonadal functions. Recent Prog. Horm. Res., 35:301-342.

Falck, B., 1959. Site of production of oestrogen in rat ovary as studied in micro-transplants. Acta Physiol. Scand., 47, Suppl. 163:1-101.

Guraya, S.S., 1976. Recent advances in the morphology, histochemistry, and biochemistry of steroid-synthesizing cellular sites in the nonmammalian vertebrate ovary. Int. Rev. Cytol., 44:365-409.

Hamberger, L., Hillensjo, T. and Ahrèn, K., 1978. Stereoidogenesis in isolated cells of preovulatory rat follicles. Endocrinology, 103:771-777.

Huang, E.S. and Nalvandov, A.V., 1979. Synthesis of sex steroids by cellular components of chicken. Biol. Reprod., 20:454-461.

Kagawa, H., Young, G. and Nagahama, Y., 1982a. Estradiol-17β production in isolated amago salmon (*Oncorhynchus rhodurus*) ovarian follicles and its stimulation by gonadotropins. Gen. Comp. Endrocrinol., 47:361-365.

Kagawa, H., Young G., Adachi, S. and Nagahama, Y., 1982b. Estradiol-17β production in amago salmon (*Oncorhynchus rhodurus*) ovarian follicles: Role of the thecal and granulosa cells. Gen. Comp. Endocrinol., 47:440-448.

Kagawa, H., Young, G. and Nagahama, Y., 1983. Relationship between seasonal plasma estradiol-17β and testosterone levels and *in vitro* production by ovarian follicles of amago salmon (*Oncorhynchus rhodurus)*). Biol. Reprod., 29:301-309.

Lance, V. and Callard, I.P., 1978. Hormonal control of ovarian streroidogenesis in nonmammalian vertebrates. In: R.E. Jones (Editor), The Vertebrate Ovary, Plenum Press, New York/London, pp. 361-406.

Nagahama, Y., Kagawa, H. and Young, G., 1982. Cellular sources of sex steroids in teleost gonads. Can. J. Fish. Aquat. Sci., 39:56-64.

Ozon, R., 1972. Estrogens in fishes, amphibians, reptiles and birds. In: D.R. Idler (Editor), Steroids in Nonmammalian Vertebrates, Academic Press, New York, pp. 390-413.

Peluso, J.J. Brown, I. and Steger, R.W., 1979. Effects of cyproterone acetate, a potent antiandrogen, on the preovulatory follicle. Biol. Reprod., 21:929-936.

Richards, J.S., Ireland, J.J., Rao, M.C., Bernath, G.A., Midgley, A.R., and Reichert, L.E., 1976. Ovarian follicular development in the rat: Hormone receptor regulation by estradiol, follicle stimulating hormone and luteinizing hormone. Endocrinology, 88:1562-1570.

Young, G., Kagawa, H. and Nagahama, Y., 1982a. Inhibitory effect of cyanoketone on salmon gonadtropin-induced estradiol-17β production by ovarian follicles of the amago salmon (*Oncorhynchus rhodurus*) *in vitro*. Gen. Comp. Endocrinol., 47:357-360.

Young, G., Kagawa, H. and Nagahama, Y., 1982b. Secretion of aromatizable Δ^4 androgens by thecal layers during estradiol-17β production by ovarian follicles of amago salmon (*Oncorhynchus rhodurus*) *in vitro*. Biomedical Res., 3:369-667.

Young, G., Kagawa, H. and Nagahama, Y., 1983. Evidence for a decrease in aromatase activity in the ovarian granulosa cells of amago salmon (*Oncorhynchus rhodurus*) associated with final oocyte maturation. Biol. Reprod., 29:310-315.

SEX CONTROL IN PACIFIC SALMON

Implications for Aquaculture and Resource Enhancement

Edward M. Donaldson[1] and George A. Hunter[1]

While techniques for controlling sex differentiation in teleosts have been investigated for some years now (Yamamoto, 1969; Hunter and Donaldson, 1983), it is only in the last decade that attention has been turned to the feasibility of applying sex control techniques to commercially important northern temperate species, in particular the salmonids. In recent years significant progress has been made at the West Vancouver Laboratory of the Fisheries Research Branch toward the development of sex control techniques for Pacific salmonids (Goetz et al., 1978, 1979; Stoss and Donaldson, 1980; Donaldson and Hunter, 1981a,b; Hunter and Donaldson, 1981; Donaldson and Hunter 1982a,b; Hunter et al., 1982a,b; Refstie et al., 1982; Hunter and Donaldson, 1983a,b; Hunter et al., 1983; Solar et al., 1983; Stoss et al., 1980). In particular, we have expanded our pilot scale studies on coho salmon to the production level and have initiated pilot or production scale studies on chum, chinook and kokanee salmon. In this brief review the objectives, techniques and current status of our studies on the production of sterile and female groups of salmon for both ocean release and aquaculture will be summarized.

Objectives

The objectives of sex control are divided into two distinct categories, those relating to the technique of sterilization and those relating to the technique of feminization. The technique of sterilization would in general only be applied to relatively abundant stocks or for aquaculture. The feminization technique also has application to abundant stocks but is particularly applicable to less abundant and even endangered stocks or to aquaculture.

Objectives for production of sterile salmonids for release:

- Redistribute harvest from hatchery to fishery by preventing anadromous migration.
- Eliminate precocious males which do not undergo their full growth potential.
- Produce larger fish by extending the life span in both ocean released and landlocked freshwater strains.

[1] West Vancouver Laboratory, Fisheries Research Branch, Department of Fisheries and Oceans, Canada

- Maintain silver bright quality year round in resident stocks and prolong availability to the fishery.
- Attempt to raise value of wide ranging species such as chum which deteriorate in quality as they approach sub-terminal and terminal fisheries.
- Provide a large buffer of cultured fish to reduce exploitation of wild stocks without having a genetic impact through interbreeding.

Objectives for production of sterile salmonids for mariculture and aquaculture:

- Eliminate problem of sexual maturation in males (all species) and females (some species) and thus:
  a) Permit harvest at a larger size and throughout the year.
  b) Maintain silver bright quality.
  c) Eliminate losses associated with males maturing in sea water

Objectives for production of all-female salmonid stocks for release:

- Enhance sub-optimal or endangered stocks by increasing the proportion of females in a given escapement.
- Permit reduction of escapement in healthy stocks while maintaining egg take.
- Reduce numbers of precocious males (jacks).
- Change the age structure of the population toward older fish.
- Increase the valuc of the commercial catch by providing a higher proportion of roe bearing fish.

Objectives for production of female salmonids for mariculture and aquaculture:

- Eliminate problem of precocious males and thus:
  a) Permit harvest at large size and over a longer period.
  b) Eliminate mortality associated with males maturing in sea water.
- Reduce costs of captive broodstock by rearing mainly females.

Techniques

There are three main techniques which have been developed or are being developed for the sterilization of salmonids.

- Androgen treatment during alevin and early feeding stages.
- Induction of triploidy by heat shock, cold shock or hydrostatic pressure treatment shortly after fertilization.
- Surgical removal of gonads.

Of these three techniques only the first has been developed to the point where it can be feasibly applied to salmon for ocean release. The advantages of androgen treatment during the alevin and early feeding stages are a) high treatment success rate, b) no treatment induced mortality, c) the procedure can be scaled up without problem. Sterile salmon live longer and grow larger than normal salmon although their actual growth rate as adults is somewhat lower than normal salmon. This is probably a result of the absence of anabolic steroid hormones which are normally produced by the testis and ovary.

Technique two, the induction of triploidy is currently in the development stage. It appears to cause effective sterilization only in female salmon and therefore should only be applied on a production basis to all female stocks. Current techniques for the induction of triploidy cause some mortality at the embryonic stage and scale up involves careful control of treatment parameters.

The third sterilization technique, surgical removal of the gonads, is the oldest of the three techniques and is effective if performed carefully, however, it is time consuming and therefore costly. It is also difficult to perform on small salmon, e.g., presmolts.

There are three types of techniques which can be used for the production of all-female groups of salmon.

- Indirect production of normal females by utilization of milt from genotypic females converted to phenotypic males by androgen treatment during the alevin and early feeding stages.
- Direct production of phenotypic females by estrogen treatment in the alevin and early feeding stages.
- Gynogenetic production of females by use of irradiated milt and temperature shock or hydrostatic pressure treatment.

Of these three techniques both the first and the second can and have been used in our studies on Pacific salmon. However, the first technique has more general application. The third technique, gynogenesis, is the subject of experimental studies and is not directly applicable to salmon for ocean release.

The advantages of the first technique are a) it is fully effective and b) the resultant fish are normal genotypic females and can be used as broodstock. The only disadvantage is that it is a two stage process requiring lead time. First female alevins are treated with androgen to produce phenotypic males with a female genotype. ''Female'' milt from these males is then used to fertilize normal production ova. The resultant offspring are all normal females.

The advantages of the second technique, which involves direct feminization using 17β-estradiol, the estrogen which occurs naturally in salmon are a) it is fully effective, b) the females produced are phenotypically indistinguishable from normal females and c) there is no treatment mortality at the optimal dosage. When using this direct feminization technique for ocean released fish it is important to use an external mark to avoid the use of returning adults as broodstock. The reason for this is that half of the returning adult females are genotypic males. The use of eggs from these genotypic males would result in a higher proportion of males in the next generation.

Application of Techniques

Sterilization

Coho salmon *Oncorhynchus kisutch*: The first sterilization of Pacific salmon on a pilot production basis was conducted on 1978 brood coho in early 1979 at Capilano Salmon Hatchery with the cooperation of Mr. Eldon Stone and Dr. Keith Sandercock. The sterile smolts were released in the spring of 1980. No jacks returned to the hatchery in the fall of 1980 and only a few incompletely sterilized fish returned in the fall of 1981. Sterile salmon were caught in the sports and commercial fishery during the summer of 1981 and again in small numbers in the summer of 1982. In 1983 a small number of tagged 5-year-old steriles have been reported from the fishery but the 1983 tag returns are incomplete at this time. The four-year-old sterile salmon were approximately twice the weight of the three-year-old sterile salmon and the five-year-olds appear to be larger again. Thus in this pilot production test the objectives of a) redistribution of harvest from hatchery to fishery and b) production of larger size salmon in the fishery were both met. As a consequence a second group of sterile coho were released from Capilano Salmon Hatchery in the spring of 1983 and a third release will take place in the spring of 1984. In addition to these releases from Capilano, releases are also planned from the Big Qualicum Hatchery.

Chum Salmon *Oncorhynchus keta*: Sexual maturation in Pacific salmon during the anadromous migration results in a significant loss in quality and value. This is particularly true in the chum salmon where value can vary by five fold between silver bright fish and fish harvested in terminal fisheries at a mouth of the natal river. In an attempt to raise the value of chum salmon caught in coastal waters we have initiated experimental and pilot studies on the sterilization or partial sterilization of chum salmon. The first release of treated chums took place at Thorton Creek Hatchery near Ucluelet in the early spring of 1983; a second release will take place at the same location in early 1984. These pilot releases from the West Coast of Vancouver Island are strictly experimental, as completely sterilized chums may not migrate close enough to shore to enter the traditional chum fishery. The sterilization or partial sterilization of chums from an "inside" stock may offer a greater chance of success, however, we would be faced with the logistical problem of releasing sufficient marked fish to obtain a significant recovery in the sequence of chum fisheries that correspond to the anadromous migratory path of the returning chums.

Kokanee salmon *Oncorhynchus nerka*: Kokanee salmon have been successfully sterilized on an experimental basis and in collaboration with the province of British Columbia field trials have been planned to determine whether the sterilization technique can be used to improve the quality and size of kokanee for the fresh-water sports fishery.

Production of All-Female Stocks

Coho salmon *Oncorhynchus kisutch*: Direct feminization with estrogen provides a means of maximizing roe production in the commercial fishery and also eliminates jacks. In the first pilot release study of this type, 1978 brood coho were feminized in early 1979 in parallel with the sterilization study described above. The treated fish were fin clipped and released in the spring of 1980. No jacks returned to the hatchery from the treated group in the fall of 1980. Substantial numbers of the all-female group were harvested by the fishery in the summer of 1981, and in the fall of 1981 more than 500 females returned to the hatchery. There were no males in the treated group and the fish were indistinguishable from normal females in ovarian size and body weight. This direct technique could be effectively applied to chums in periods when the roe market is strong.

Chinook salmon *Oncorhynchus tschawytscha*: It is widely recognized that chinook stocks are currently depressed and that only two of the many hatcheries rearing chinooks in British Columbia have sufficient adult females returning to them to meet their egg requirements. This problem of low hatchery escapements is exacerbated by the fact that the female:male ratio in returning brood stock has dropped to only 1:4 or lower at a number of facilities. Two of the contributing factors here are the lower average age at maturity in male chinook and the fact that older chinook migrate further north along the Pacific coast and are thus exposed to greater fishing pressure. Improved management of the fishery on an international basis used in combination with this newly developed biotechnology to improve the female:male ratio in chinooks could if implemented soon enough result in a reversal in the current decline in chinook stocks.

The development of techniques at the West Vancouver Laboratory for the production of chinook milt that contains only female chromosomes has resulted in the production of normal all-female chinook on an experimental basis. In the fall of 1983, a significant batch of chinook eggs transplanted from the Big Qualicum Hatchery were fertilized at the Capilano Salmon Hatchery with "all-female" milt produced at the West Vancouver Laboratory. The resultant all-female smolts will be released in the spring of 1984 and will return as female broodstock in 1986 and successive years.

In order to produce larger quantities of female milt, chinook alevins and fry were treated with androgen at Capilano hatchery in early 1983 and were released as all-male smolts in the spring of 1983. These externally marked phenotypic males will return to Capilano hatchery as 2-, 3- and 4-year-olds in the fall of 1984 and successive years. The milt from these special males will be used to fertilize production eggs and will result in the release of smolts that are 75% female. The basis for this percentage is that half of the alevins and fry treated with androgen would have been males regardless of treatment and thus would produce milt in a 50:50 male:female ratio, while the other half of the treated alevins would be genotypic females and would produce 100% female milt. It is hoped to extend the use of this technique to other facilities where chinook broodstock are in short supply.

Future Directions

The biotechnology of controlled sex differentiation in Pacific salmon is still in its infancy. The studies that have been conducted to date, however, clearly show that we now have at our disposal or under development powerful new tools for the management and enhancement of our salmon resource.

All pilot production scale releases into the ocean require careful evaluation of a) geographic distribution, b) contribution to specific fisheries and gear types, c) contribution to the fishery in successive years at successive ages and d) migratory patterns. In addition the release of a large buffer of sterile salmon in combination with limitation of fishing effort could be explored as a means of reducing the impact of the fishery on wild stocks. This buffer of sterile fish would remain in the fishery rather than maturing and returning to the point of release. In such a project it would be necessary to design studies to test for possible displacement of resident stocks to outside waters and to test for predatory impact on juveniles of salmonid and non-salmonid species.

With regard to laboratory experimentation on sex control for ocean release it is important that further development work be conducted on the improvement and where possible simplification of existing techniques for individual target species. In addition, further studies on triploid production and gynogenesis are warranted to establish the feasibility of future application.

References

Donaldson, E.M., and Hunter, G.A., 1981a. Production of all-female and sterile stocks of Pacific salmon for ocean ranching. ORI Symp. on Fish Migration and Reproduction, Tokyo, Dec. 1-2. Abstr. p. 29.

Donaldson, E.M., and Hunter, G.A., 1981b. Control of sex ratio in Pacific salmon broodstock. In: T. Nosho (Editor), Salmonid Broodstock Maturation. Proceedings of a workshop held at Seattle, Washington, May 20-22, 1980, and March 11, 1981. A Washington Sea Grant Publ., Univ. Washington, Seattle. p. 80.

Donaldson, E.M., and Hunter, G.A., 1982a. Sex control in fish with particular reference to salmonids. In: W.S. Hoar (Editor), Symposium, Fish Reproduction: Behavior and Endocrinology. Can. J. Fish. Aquat. Sci. 39:99-110.

Donaldson, E.M., and Hunter, G.A., 1982b. The ocean release and contribution to the fishery of all-female and sterile groups of coho salmon (*Oncorhynchus kisutch*). In: H.J.Th. Goos and J.J. Richter (Editors), Proceedings of the International Symposium on Reproductive Physiology of Fish, Wageningen, The Netherlands, 2-6 August, 1982, p. 78.

Goetz, F.W., Donaldson, E.M., Hunter, G.A., and Dye, H.M., 1978. The effects of estradiol-17β and 17α-methyltestosterone on early gonadal differentiation in the coho salmon *Oncorhynchus kisutch*. In: P.J. Gaillard and H.H. Boer (Editors), Comparative Endocrinology. Elsevier/North Holland, Amsterdam. pp. 92-93.

Goetz, F.W., Donaldson, E.M., Hunter, G.A., and Dye, H.M., 1979. Effects of estradiol-17β and 17α-methyltestosterone on gonadal differentiation in the coho salmon, *Oncorhynchus kisutch*. Aquaculture 17:267-278.

Hunter, G.A., and Donaldson, E.M., 1981. Experimental production of all female and sterile groups of coho salmon (*Oncorhynchus kisutch*). Proceedings of the 31st Annual Northwest Fish Culture Conference, Courtenay, B.C., Dec. 2-4, 1980. 234-241.

Hunter, G.A., and Donaldson, E.M., 1983a. Hormonal sex control and its application to fish culture. In: W.S. Hoar, D.J. Randall and E.M. Donaldson (Editors), Fish Physiology, Vol. IX, Part B, Behavior and Fertility Control, 5:223-303. Academic Press, New York.

Hunter, G.A. and Donaldson, E.M., 1983b. Contribution of ocean released sterile coho to the commercial and sports fishery. Int. Symp., Salmonid Reproduction, Bellevue, Wash., Oct. 31-Nov. 2, 1983, p. 30-31.

Hunter, G.A., Donaldson, E.M. and Stone, E.T., 1982. Control of sex in salmonids; pilot release of sterile and all-female groups of coho salmon *Oncorhynchus kisutch* from the Capilano Salmon Hatchery. Proceedings of the 32nd Annual Northwest Fish Culture Workshop, Vance Tyee, Tumwater, Washington, December 1-3, 1981. p. 83-87.

Hunter, G.A., Donaldson, E.M., Goetz, F.W., and Edgell, P.R., 1982. Production of all female and sterile groups of coho salmon (*Oncorhynchus kisutch*) and experimental evidence for male heterogamety. Trans. Am. Fish. Soc. 111:367-372.

Hunter, G.A., Donaldson, E.M., Stoss, J., and Baker, I., 1983. Production of monosex female groups of chinook salmon (*Oncorhynchus tshawytscha*) by the fertilization of normal ova with sperm from sex reversed females. Aquaculture 33:355-364.

Refstie, T., Stoss, J., and Donaldson, E.M., 1982. Production of all female coho salmon (*Oncorhynchus kisutch*) by diploid gynogenesis using irradiated sperm and cold shock. Aquaculture 29:67-82.

Solar, I.I., Donaldson, E.M., and Hunter, G.A., 1983. Sex control in rainbow trout for mariculture in British Columbia. Int. Symp., Salmonid Reproduction, Bellevue, Wash., Oct. 31-Nov. 2, 1983, p. 24.

Stoss, J., and Donaldson, E.M., 1980. Moglichkeiten zur Geschlechtskontrollebei Salmoniden. (Possibilities of sex control in salmonids). Fischer und Teichwirt 12:356-359.

Yamamoto, T., 1969. Sex differentiation. In: W.S. Hoar and D.J. Randall (Editors), Fish Physiology, 3:117-175. Academic Press, New York.

ABSTRACTS–Endocrinology

The Application of Radioimmunoassays for Sex Steroids, Gonadotropin, and Vitellogenin to the Study of the Reproductive Cycle of Rainbow Trout *Salmo gairdneri*

A. P. Scott, J. P. Sumpter, S. M. Baynes, and P. R. Witthames, Ministry of Agriculture, Fisheries, and Food, Lowestoft, United Kingdom*
** Brunel University, Middlesex, United Kingdom*

In rainbow trout females, significant changes occur in plasma levels of sex steroids, gonadotropin and vitellogenin during the reproductive cycle. While much attention has previously been directed to the levels of these hormones during advanced stages of maturity (i.e., exogenous vitellogenesis and ovulation), very little has been directed to the changes that occur during the very early stages of gonadal growth (i.e., the beginning of the secondary oocyte growth phase).

We present data on plasma hormone and vitellogenin levels in two-year old, virgin female rainbow trout over the months from January to June. This period encompasses the initiation of secondary oocyte growth phase as assessed histologically.

Data are also presented on hormone levels in triploid female rainbow trout. These fish are effectively sterile (the ovaries contain only oogonia), so the plasmas constitute a baseline for the present investigations which are concerned with small changes in hormone levels.

Plasma Steroid Profiles during Sexual Maturation in Salmonids

Kouhei Yamauchi, Hokkaido University, Hakodate, Japan

Plasma levels of various steroids were examined at different stages of sexual maturity of three species of salmonids (amago salmon, *Oncorhynchus rhodurus*; masu salmon, *O. masou*; chum salmon, *O. keta*). Estradiol-17β (E_2), testosterone (T), 11-ketotestosterone (11-ketoT) and 17α, 20β-dihydroxy-4-pregnen-3-one (17α, 20β-diOHprog) were measured by radioimmunoassay. Plasma steroid hormone profiles in three species of salmonids were similar and will be treated together. In females, E_2 levels gradually increased during vitellogenesis and peaked at the end of the vitellogenic period and declined markedly in mature and ovulated fish. T levels lagged behind and followed that of E_2 during vitellogenesis, but remained high in mature and ovulated females. 17α, 20β-diOHprog levels were low during the major part of the preovulatory period and rapidly elevated in mature and ovulated females. Detailed changes in the plasma levels of E_2 and 17α, 20β-diOHprog levels were investigated in masu salmon during the course of final maturation and ovulation. E_2 levels were high at the beginning of the experiment (8 days prior to ovulation) and declined to basal levels 4–6 days prior to ovulation. In contrast, 17α, 20β-diOHprog levels were either not detectable or low when E_2 levels were high and rapidly increased after E_2 levels declined to basal levels, reaching a peak 2–4 days prior to ovulation. The levels gradually decreased thereafter.

In males, plasma levels of 11-ketoT increased during rapid testicular development, followed by a sharp drop during spawning period. A similar pattern was observed for T although T was consistently lower than 11-ketoT. 17α, 20β-diOHprog levels were very low during the major part of the development of the testis, and rapidly increased at the onset of spermiation; the levels remained high during the period of active spermiation and sharply declined thereafter. The possible roles of steroid hormones in gamete development and maturation in salmonids will be discussed.

Evolution of the Follicular Sensitivity *in vitro* to Maturation-inducing Hormones at the End of Vitellogenesis in Rainbow Trout *Salmo gairdneri:* Role of Estradiol-17β

Bernard Jalabert and Alexis Fostier, Institut National de la Recherche Agronomique, Laboratoire de Physiologie des Poissons, Rennes Cedex, France

Trout ovaries exhibiting oocytes at a defined stage around the end of vitellogenesis (end of vitellogenesis—GV −, subperipheral germinal vesicle—GV ±, peripheral germinal vesicle—GV + were processed for *in vitro* incubations in standard conditions, in order to examine possible relationships between the following parameters: oocyte sensitivity to the maturing steroid 17α-hydroxy-20β-dihydroprogesterone (17α, 20β-OH-P), follicular sensitivity to the maturational gonadotropin s-GtH inhibitory potency of exogenous estradiol-17β (E_2) over s-GtH-induced maturation, and level of E_2 in the plasma of donor fishes. The sensitivity to hormones was estimated by the median efficient dose (MED) for morphological oocyte maturation *in vitro*.

Globally, the follicular sensitivity and the oocyte sensitivity exhibit a significant covariation. However, when attention is paid to the oocyte morphological stage, the peripheral migration of the GV appear to coincide with a jump in oocyte sensitivity to 17α, 20β-OH-P.

A significant correlation can be observed between plasma E_2 level and the follicular sensitivity to s-GtH, but not between plasma E_2 and the oocyte sensitivity to 17α, 20β-OH-P. The inhibitory effect of E_2 on s-GtH-induced maturation *in vitro* appears all the more important as the follicular sensitivity is high, and can be demonstrated even with low doses, in the physiological range (0.05 μg/ml). Taken together these data lead to the conclusions that the peripheral migration of the GV is a morphological event which appears to coincide statistically with a jump in oocyte sensitivity and that E_2 is an important physiological regulator of follicular sensitivity to GtH.

Ovulatory and Steroidal Responses in Coho Salmon and Steelhead Trout Following Administration of Salmon Gonadotropin and D-Ala^6, Des-Gly^{10} Ethylamide(GnRHa)

S. A. Sower, University of New Hampshire; R. N. Iwamoto, W. W. Dickhoff, and A. Gorbman, University of Washington

Hormone injection experiments were conducted for two years during the fall and early winter when adult coho salmon and steelhead trout, respectively, spawn. Successive injections of partly purified coho salmon gonadotropin and synthetic analogue of hypothalamic GnRHa advanced ovulation by two to three weeks in coho salmon and steelhead trout in all tests. In these same tests, all treatments of salmon gonadotropin followed by one of 4 doses of GnRHa (50, 5, 0.5, or 0.05 micrograms/kilogram) in coho salmon significantly depressed estradiol levels whereas androgen levels were significantly elevated compared to controls. Treatment with salmon gonadotropin followed by only one dose of GnRHa (60 microgram/kg) in steelhead trout caused a similar response of a significant decrease of estradiol. In summary, the doses of GnRHa (5, 0.5, or 0.05 microgram/kg) following injection of coho salmon gonadotropin are the lowest doses that have been reported to be effective for acceleration of ovulation in salmonids. Furthermore, part of the mechanism of normally accelerated ovulation is evident by a decrease in estradiol levels concomitant with an increase in androgen levels.

Biochemical Changes Occurring in Female Salmon *(Salmo salar)* During Vitellogenesis

Carl Haux, University of Göteborg, Göteborg, Sweden; and Ulf-Peter Wichardt, Salmon Research Institute, Älvkarleby, Sweden

During a period of the reproductive cycle in female fish, a specific lipophosphoprotein, vitellogenin (VG), is synthesized in the liver, transported in the blood to the ovaries and incorporated in the oocytes. Estradiol-17β appears to be the main steroid involved in the induction of VG synthesis. This is supported by findings in adult female Atlantic salmon, where a correlation between plasma VG and estradiol-17β has been reported (Idler et al., Can. J. Fish Aquat. Sci. 1981, 38, 405-413). Since VG is a large multicomponent protein, its appearance in the blood alters the biochemical plasma composition dramatically.

The object of the present study was to follow changes in some plasma parameters that are related to vitellogenesis by repeated blood sampling on female salmon undergoing sexual maturation and spawning. Increased knowledge of these processes could have applications in brood fish handling, where environmental factors, such as temperature, handling stress, infections, prophylactic treatments etc., may influence vitellogenesis and subsequently the production of eggs and fry.

In the present study, females were trapped near Älvkarleby on their spawning migration in the river Dalälven during July. Totally eight females were tagged and repeatedly sampled roughly every second week, starting in August and ending in December. All females were stripped of eggs within one week in early November.

The amount of plasma VG, measured as phosphate bound to protein (PP), was about 2.4 mM in August, increased to 5.7 mM in September, and decreased successively to 2.4 mM at the time of spawning. The level of PP decreased then further to 1.3 mM in December, which is close to the amount of PP present in adult male salmon (approximately 1 mM). Elevated levels of the total plasma protein content (PR), which reached a maximum of 6.8 g/100 ml in late August and decreased to 4.5 g/100 ml at spawning, presumably reflected the presence of VG in the plasma. This assumption is further supported by a correlation between plasma PP and PR ($r = 0.54$). Similarly, total plasma calcium (CA), and to a lesser extent, total plasma magnesium (MG) followed the increase and decrease in plasma VG closely: CA peaked at 4.7 mM in September and MG at 1.3 mM in late August, were reduced to 3.2 and 0.9 mM respectively during spawning, and reached normal values of 2.4 and 0.8 mM in December. The increase in total plasma CA as well as the close correlation between plasma PP and CA ($r = 0.82$), indicate that salmon VG binds CA in a similar way to that observed in other species. Possibly, also MG could bind to VG ($r = 0.44$). During the most intense period of vitellogenesis, the amount of plasma phospholipids (PL) were markedly elevated (by 30 to 50%). This increase could probably be attributed to VG, which usually contains both triglycerides and PL.

Reproductive Activity of a Twice-Annually Spawning Strain of Rainbow Trout

Katsumi Aida, Show W. Lou, Isao Hanyu, University of Tokyo; Kiyoshi Sakai, Minoru Nomura, Tokyo University of Fisheries; Mikio Tanaka, Shiro Tazaki, Fisheries Laboratory of Saitama Prefecture, Japan

A strain of rainbow trout that breeds twice a year was found at Kumagaya Branch of Fisheries Laboratory of Saitama Prefecture. At present after several selections, almost all females of the strain ovulate in the ordinary breeding season from November to January and 60–70% of the females ovulate again in another breeding season from April to July. It is still uncertatin why this strain can breed at intervals of about six months under almost constant temperatures and natural photoperiods. In this investigation, we studied the endocrine background of reproductive activity of this strain.

Ninety four females which ovulate in November were tagged. Blood and a piece of ovary were repeatedly sampled for a year. Ovulation occurred from April to July (62.3%) out of 77 females and from November to January in 59 (96.7%) out of 61 females which survived.

In 48 females that ovulated from April to July, plasma estradiol-17β (E_2) began to increase in January, attaining highest values from March to May. From January to March plasma GTH of these fish was significantly higher than that of the females which failed in vitellogenesis in spring. On the other hand, in the non-vitellogenic females, plasma E_2 and GTH showed small increase from January to February, but decreased in March. Lengthening day length in spring probably inhibited GTH secretion in this group, but not in the fish which had already started the vitellogenesis by the criticial period in spring.

In almost all females, synchronous increase of plasma E_2 was induced in July, probably by shortening day length. E_2 attained highest values in October and ovulation synchronously occurred in December.

Now in Japan, the spawning of rainbow trout mainly occurs from September to November as a result of the repeated selection of early spawners. The selection is probably responsible for this twice-annually spawning strain of rainbow trout.

Endocrine and Ovarian Changes in Three Strains of Rainbow Trout Subjected to Both Constant and Seasonally Changing Photoperiods

J. A. K. Elliott, N. Bromage, J. Springate, University of Aston, Birmingham, United Kingdom

Although it is ultimately the environment and more particularly the photoperiod that is involved in the overall control of reproduction in trout, it is now well recognized that it is changes in the hormonal milieu that directly control the sequence and timing of the different phases of ovarian development and maturation such that spawning occurs at the most propitious time of year. However, few studies have investigated whether the spawning separation shown by different strains of salmonids are due to a response to photoperiod cues of different lengths or to a modification in the endocrine control. The present study investigates this question by comparing the endocrinological and histological development of three strains of rainbow trout subjected to three types of seasonal and constant photoperiod regime.

The three strains used in this study were: Caribou (November spawning), Grampian (December) and Whitebrook (January). Over a period of 28 months the same groups of fish were subjected to the following three photoperiod cycles:

(1) Constant long days (18L:6D), followed by a direct switch to short days on June 21st

(2) A compressed 6-month seasonal photoperiod

(3) A normal 12-month seasonal photoperiod, 6 months out of phase

Groups of fish from each strain were blood sampled regularly and serum assayed for estrone, estradiol-17β, testosterone, vitellogenin and calcium. Fish from the same stocks were sacrificed for histological analysis of the ovary and for determinations of GSI, HSI and mean oocyte diameter. The photoperiod was provided by fluorescent lighting controlled by a 24-hour timeclock and water temperature (10°C) and feeding (0.5% bw) kept constant.

Under all photoperiod cycles female fish of the three strains showed similar sequences of endocrine changes and ovarian development. However, the spawning identity of the three strains was maintained under all regimes, although the timing of the speed of attainment of the various serum profiles was clearly modified to account for the perceived differences in spawning time. Initial increases in serum levels of estrone and estradiol-17β were followed by increases in vitellogenin, calcium and testosterone. Similar, although much reduced, changes were observed in immature fish of one strain at a time of year suggestive of a practice or dummy run, one year before the first spawning, as a rehearsal for full reproductive development.

These data demonstrate that different strains maintain their separate spawning times by responding differentially to photoperiodic cues rather than by any modification of their endocrine control. Collectively the responses suggest that the different strains respond to different numbers of daily photoperiodic stimulations, rather than photoperiodic cues of different lengths.

Induction of Ovulation in Atlantic Salmon with Pelleted LHRH Analog

L. W. Crim, Memorial University of Newfoundland, St. John's Canada; and B. D. Glebe, North American Salmon Research Center, St. Andrews, New Brunswick, Canada

Spawning in Atlantic salmon begins in the late fall and it may require several weeks for completion in all individuals of some stocks. Inducing spawning with LHRH or LHRH analogs (LHRH-A) may become the method of choice since these simple peptide hormones are commercially available at competitive prices relative to preparations of fish gonadotropic hormone. The objectives of our 1982 studies included an examination of the value and practicality of using pelleted, long-acting LHRH-A for advancing and synchronizing spawning of the female Atlantic salmon. Two separate experiments were conducted with sea cage-reared salmon beginning on September 14 and October 4, respectively. Females were treated with intraperitoneal implants of pelleted LHRH-A or control fish received the sham surgical procedures alone. The fish were frequently checked for signs of ovarian maturity and weekly blood samples were collected until the time of spawning. On the day of spawning, females were stripped and the eggs were fertilized with milt obtained from two males. The eggs were incubated in individual batches and fertility and viability to the eyed stage were recorded. In September ovary development was variable ranging from 5.9–19.7% GSI in a small group of initial controls. Plasma gonadotropic hormone (GtH) levels were elevated in females treated with pelleted LHRH-A but spawning was advanced (October 6) in only 3 (20% of group) fish. In experiment 2 beginning October 4, LHRH-A treatment increased plasma GtH and also accelerated and synchronized ripening of the treated females (94% of group) within 11 days. Compared with sham control females LHRH-A advanced maturation of females approximately 3 weeks. Whereas egg quality was extremely poor in the September trial, egg fertility and viability were acceptable for the LHRH-A stimulated fish in October. We conclude that pelleted LHRH-A can be successfully used to advance maturation and ovulation in Atlantic salmon, especially during the later phases of ovarian development. LHRH-A treatment also appears capable of synchronizing ovulation in female Atlantic salmon.

The Use of Des-Gly^{10} (D-Ala^{6}) LH-RH-Ethylamide to Induce Precocious Ovulation in Adult Female Coho Salmon *(Oncorhynchus kisutch)*

M. S. Fitzpatrick, D. M. Oberbillig, B. K. Suzumoto, and C. B. Schreck, Oregon Cooperative Fishery Research Unit, Oregon State University, Corvallis*
** Oregon Aqua Foods, Inc., Springfield, Oregon*

When adult hatchery coho salmon return to seawater or freshwater facilities, they must often remain in raceways for extended periods of time before ripening. This often results in significant mortality before ovulation. To circumvent this problem, as well as to simply obtain eggs when needed, we investigated means of inducing precocious final maturation.

Adult female coho salmon held in fresh water were given intraperitoneal injections of either the mammalian luteinizing hormone-releasing hormone analog (LH-RHa) des-Gly^{10} (D-Ala^{6}) LH-RH-ethylamide at various dosages or injections of saline. The mean number of days to ovulation in all groups of fish receiving only LH-RHa was significantly ($p<.01$) lower than that in saline injected controls. Within 10 days of the initial injection, more than half of the fish in groups receiving single injections of LH-RHa at various dosages had ovulated compared to 11% of the controls. Within two weeks of the initial injection, 85% of the fish in groups receiving two injections of LH-RHa at various dosages had ovulated while less than 50 percent of the controls had ovulated. This demonstrates that treatment of female coho salmon with LH-RHa can cause significant acceleration of final maturation leading to precocious ovulation.

In order to better understand the biology of salmon, our lab has also been involved in the development of endocrine profiles for adult fish. We are investigating the dynamics of various sex steroids in the plasma of salmon throughout the spawning run. Preliminary results from coho salmon and chum salmon *(O. keta))* indicate trends in the plasma titers of progesterone, estradiol, and 17α-hydroxy-20β-dihydroprogesterone which may elucidate the necessary criteria for the actions of maturation-inducing agents.

Changes in Plasma Estradiol-17-β and 17-α, 20-β Dihydroxy-4-Pregnen-3-One during Spontaneous and Induced Ovulation in Coho Salmon

G. Van Der Kraak, University of British Columbia and E. M. Donaldson, Department of Fisheries and Oceans, West Vancouver, British Columbia, Canada

Current evidence suggests that in salmonids, vitellogenesis and oocyte maturation are mediated by estradiol-17β and 17α, 20β dihydroxy-4-pregnen-3-one (17α, 20βP) respectively. Few studies have examined the switch from estradiol-17β to 17α, 20βP production during the preovulatory period. In the present study, estradiol-17β and 17α,20βP were measured in plasma samples taken during spontaneous or gonadotropin induced reproductive activity to determine the timing of steroid changes in relation to oocyte maturation and ovulation. Spontaneous reproductive activity was characterized by a rapid decline in plasma estradiol-17β commencing 12 days prior to ovulation and a large increase in plasma 17α, 20βP six days prior to ovulation. This preovulatory increase in 17α, 20βP coincides with the time of germinal vesicle breakdown. Fish injected with des-Gly^{10}DAla^{6}LH-RH-ethylamide alone or in combination with SG-G100 showed a biphasic increase in plasma 17α, 20βP levels. While plasma 17α, 20βP levels were elevated within three hours of injection, the surge associated with oocyte maturation was delayed four to eight days. In contrast, estradiol-17β levels while unchanged for 24 hours, declined to basal levels within 6 days. These results suggest that salmon have 20β hydroxysteroid dehydrogenase at least one month prior to the expected time of ovulation. It is not known if the concentration of this enzyme is low, necessitating the synthesis of additional enzyme or alternatively, the removal of an inhibitory factor to produce the high levels of 17α, 20βP associated with oocyte maturation. Analysis of plasma estradiol-17β and 17α, 20βP during induced ovulation indicates that the decline of estradiol-17β consistently precedes the surge in 17α, 20βP suggesting that estradiol-17β may function as an inhibitor of 17α, 20βP synthesis.

Advancement and Synchronization of Spawning in *Salmo gairdneri* and *S. trutta* Following Administration of Pimozide and LHRH-A

Rolland Billard, P. Reinaud, M. G. Plouidy, B. Breton, Laboratoire de Physiologie des Poissons, Institute National de la Recherche Agronomique, Jouy-en-Josas, France

The spawning of rainbow trout was advanced by more than 1 month in 60% of females after pimozide injection (10 mg/kg body weight) followed by LHRH-A injection (1 μg/kg) 6 h. later. Advancement was clearly less marked when we injected either LHRH-A or pimozide alone and when the controls were given only saline solution. Ovum fertility was comparable in all groups.

With the aim of avoiding two successive injections, we tried administering pimozide (10 mg/kg in an injection) and LHRH-A (implant containing 50 and 5 μg/kg body weight) simultaneously in brown trout during full spawning. The layings of females given pimozide plus LHRH-A were grouped over a 4-day period and those of females given only pimozide or a blank implant over 8 days.

Effects of Estradiol-17-β on Gonadal Differentiation in Two Species of Salmonids, the Masu Salmon, *Oncorhynchus masou*, and the Chum Salmon, *O. keta*

Masaru Nakamura, Faculty of Medicine, Teikyo University, Tokyo, Japan

The effect of estrogen treatment on the induction of gonadal feminization was investigated in two species of salmonids, masu and chum salmon. Histological examination reveals that the two sexes can be distinguished clearly in masu salmon 13 days after hatching and in chum salmon 25 days after hatching. Masu salmon fry were immersed in water containing estradiol-17β (E_2) at concentrations of 0.25, 0.5, 1, 2, 5, 10, 20, 50,100 and 200 μg/1 for 18 days starting from 5 days after hatching. Sex differentiation was assessed histologically at 30, 50, 90 and 360 days after hatching. Histological examination of fish at 30 days revealed that in the 0.5 to 5 μg/1 treated groups, nearly 100% fish were female with gonads which closely resembled those of female controls; retarded ovaries were observed only in one or two fish in each group. In addition, inspection of gonads of 0.5 to 5 μg/1 groups at 360 days confirmed completed feminization. Most fish of these groups were grown to maturity. On the other hand, masu salmon receiving higher doses of E_2 (10 to 200 μg/1) experienced high mortality while fish with the lowest concentration (0.25 μ/1) had intersexual gonads.

Chum salmon fry at various stages of gonadal sex differentiation were immersed in water containing E_2 at concentrations of 0.5, 1 and 2 μg/1 for a period of 15 to 67 days. Unlike the situation in masu salmon, complete feminization could not be induced in chum salmon by these treatments. However, partial feminization was observed in fish treated at the lowest concentration (0.5 μg/1) during the period from 6 to 34 days after hatching. The testes of this group consisted of small oocytes interspersed with undeveloped germ cells with stroma. These results are discussed in relation to the effective dose and period of estrogen treatment for the successful feminization of gonads in two species of salmonids.

On Modifications of Gonadotropic Cell Ultrastructure in the Intact Mature Dwarf Male *Oncorhynchus formosanus* and Juvenile *O. masou* after Injection of Salmon Gonadotropins

A. A. Maximovich, V. V. Pljusnin, and V. N. Poljakov, Laboratory of Functional Morphology of Cells, Institute of Marine Biology, Vladivostok, U.S.S.R.

The ultrastructure of gonadotropic cells (GTG-cells) of adenohypophysis in the dwarf male salmon *Oncorhynchus formosanus* during gonadal maturation and in the juvenile masu salmon *(Oncorhynchus masou)* in 6, 12, and 24 hours after a single injection of salmon gonadotropin (GTG) was studied.

Two fractions of GTG isolated from the hypophyses of mature males (fraction M_2) and females (fraction F_1) of the pink salmon *Oncorhynchus gorbuscha* and purified by column gelfiltration with Sefadex G-100 have been used in this study.

In all the fish individuals granular, vesicular and globular GTG-cells were detected. It was established that the percentage of these GTG-cells in the adenohypophysis distinctly changed during sexual maturation and after injecting the fractions F_1 and M_2 to immature salmon. Accordingly, the question is discussed as to whether the detected types of GTG-cells are different functional stages of gonadotropocytes or if they are responsible for production of different gonadotropic hormones.

Development of Atherosclerotic Lesions in Coronary Arteries of Atlantic Salmon During Sexual Maturation

R. L. Saunders, Department of Fisheries and Oceans, St. Andrews, New Brunswick; A. J. Farrell, Mount Allison University, Sackville, New Brunswick; H. C. Freeman, Department of Fisheries and Oceans, Halifax, Nova Scotia, Canada

Atherosclerotic lesions have been observed in coronary arteries of Atlantic salmon and other salmonids at the time of sexual maturation. It has been suggested that these lesions are in association with elevated levels of sex hormones. The purpose of our study is to document development of atherosclerotic lesions in Atlantic salmon in relation to the natural maturation cycle between June and October. Samples of maturing and non-maturing individuals were killed monthly for determination of blood levels of cholesterol and high density lipids as well as the hormones testosterone, 11-keto-testosterone and estradiol; heart-ventral aorta samples were taken for histological examination. The determination of sex hormone levels and histological examination of coronary arteries is in progress. Additional samples will be collected during 1983 to learn whether or not atherosclerosis is confined to maturing individuals and, if so, to study the time course of its development and related patterns of sex hormone levels.

Sex Reversal in Atlantic Salmon: Problems with High Doses of Estradiol and DES

S. A. Sower, University of New Hampshire; W. W. Dickhoff, T. Flagg, J. Mighell, University of Washington; and C. Mahnken, National Marine Fisheries Service, Seattle

Atlantic salmon were fed Oregon Moist pellets incorporated with different doses of estradiol (20 or 2 mg/kg diet) or diethylstilbesterol (DES) (10, 1 or 0.1 mg/kg diet) in an attempt to reverse the sex of the salmon from male to female during the time of sexual differentiation. Atlantic salmon treated with a high dose of estradiol at 20 mg/kg had the highest mortality compared to controls and other treatment groups. Also, the salmon treated with the highest dose of the estrogens contained the highest percentages of female fish.

It is not clear from this study, whether differential mortality occurred, causing male mortality resulting in survival of the females, or whether sex reversal had occurred. These possibilities are discussed. We also present data on the occurrence of precocial sexual development in 42% of the control males which were only six months old. Precocial sexual development did not occur in the estrogen-treated male fish. The implications are presented.

Purification Studies of Coho Salmon Pituitary Glycoprotein Hormones

P. Swanson, W. W. Dickhoff, A. Gorbman, University of Washington, Seattle

Reproductive processes in salmon may be regulated by both the thyroid and reproductive endocrine systems. Important components of these two systems are the pituitary glycoprotein hormones, gonadotropic hormone (GtH) and thyroid stimulating hormone (TSH). Research designed to examine the specific physiological roles of GtH and TSH in salmon reproduction has been limited by (1) the lack of salmon GtH which has been demonstrated to be uncontaminated with TSH and (2) the lack of pure salmon TSH. The goal of this investigation was to isolate coho salmon TSH and GtH.

Two sources of pituitary glands were used. Pituitaries were collected from sexually mature coho salmon at the time of spawning and from immature coho salmon held in seawater net pens. Chemical fractionation procedures included extraction in acid alcohol and chromatography with Sephadex G-100, Sulphopropyl Sephadex C-50, and DEAE-Sephacel. During the purification steps assays for gonadotropic and thyrotropic activities were based on the ability of test materials to elevate plasma levels of thyroxine (T_4) and estradiol 17-β after injection into underyearling coho salmon (body wt. 15-30 g). Fractions from both sources of glands contained thyrotropic and gonadotropic activities. A fraction from the pituitaries of seawater-maintained salmon had greater TSH activity than an approximately equivalent fraction from the pituitaries of spawned adult fish. Extracts of salmon pituitaries were more than twice as potent as bovine TSH (Sigma Chemical Co.) in elevating plasma T_4. A seasonal variation in responsiveness to test fish TSH and GtH fractions was observed.

Contribution of Ocean Released Sterile Coho to the Commercial and Sports Fishery

George Hunter and Edward M. Donaldson, Department of Fisheries and Oceans, West Vancouver, Canada

This experiment was designed to determine whether sterile coho salmon released from a production hatchery would remain in the ocean at the time of the normal anadromous migration, remain accessible to the commercial and recreational fisheries and continue to grow beyond the normal time of death.

Approximately 39,000 coho, administered a sterlizing androgen treatment in the spring of 1979 were marked with coded wire tags and released in the spring of 1980. Contribution to both the recreational and commercial fisheries from 1980-1983 has been monitored via the Head Recovery Program. No sterile fish have returned to the hatchery. These fish have however remained in coastal waters and have contributed to both fisheries at rates comparable to normal production fish. Further, the fish have continued to grow throughout their prolonged ocean residence. Data will be presented on the spatial and temporal distribution of the catch.

Is the Poor Survival of Lake Erie Coho Salmon Eggs Due to Nutritional Deficiency of the Eggs or to Endocrine Dysfunction in the Adult?

J. F. Leatherland and P. F. Morrison, University of Guelph; and R. A. Sonstegard, McMaster University, Hamilton, Ontario, Canada

Coho salmon (*Oncorhynchus kisutch*) were introduced into the Great Lakes in the early 1960's from stock originating in either Oregon (Lakes Michigan, Ontario, Superior and Erie) or British Columbia (Lake Ontario). In addition to the thyroid neoplasia exhibited to varying degrees by Great Lakes coho salmon, considerable interlake variation was noted in the survival of eggs to the fry stage. In order to determine whether this apparent interlake difference in survival could be attributed to differences in water conditions, husbandry methods, etc. at the different hatcheries, coho salmon eggs were collected during the peak migration of salmon from Lakes Ontario (Oregon and British Columbia stock), Erie and Michigan. They were fertilized, hardened in the local river water and transported to the University of Guelph to be reared under standardized conditions. Survival to hatch was 79%, 86%, 78%, and 24% for eggs for Lake Ontario Oregon stock, Lake Ontario British Columbia stock. Lake Michigan and Lake Erie salmon, respectively. These data generally agree with the survival values obtained by the different hatcheries, and suggest that the low survival of eggs from Lake Erie salmon is intrinsic and not due to hatchery factors. This is supported by the observation that eggs from small Lake Erie females showed a lower survival (9%) than eggs taken from larger females (26%). Total lipid, triglyceride, phospholipid and protein content were similar in eggs from the four sources of salmon, suggesting that the poor survival of the eggs was not related to nutritional deficiency. However, differences in serum gonadotropin 17β-estradiol and testosterone levels in salmon from different Great Lakes suggest that the high mortality of the Lake Erie eggs between fertlization and hatch may be related either to a dysfunction in steroid secretion or an enhanced catabolism of gonadal steroids.

Genetics

Session Leaders

Graham A. E. Gall
Trygve Gjedrem

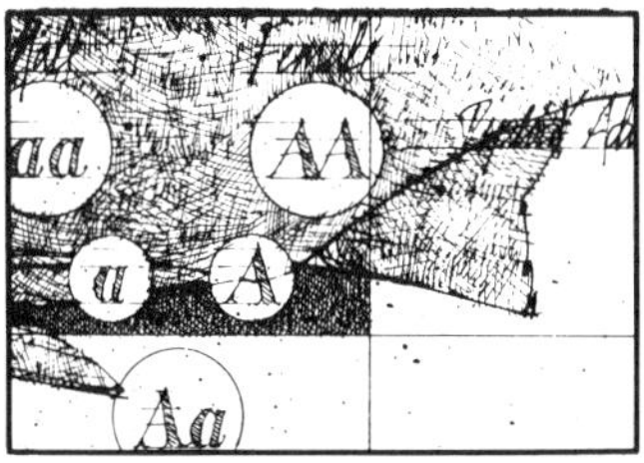

QUANTITATIVE GENETIC ASPECTS OF REPRODUCTION IN SALMONIDS

Things Are Not Always What They Seem To Be

Graham A.E. Gall[1]

Artificial propagation of salmonids is undertaken with the objective of improving the production of fish. These efforts include breeding programs most often designed to bring about changes in performance while a few are designed to maintain the status quo of given populations. The characteristics of the fish most often considered important are traits associated with growth, egg production, and age at sexual maturation. One assumption can always be made: the artificial spawning of a population will result in some type of change in the average performance of the population. These changes will occur as a response of the organism to the conditions imposed by the hatchery management protocol chosen for the program, a process referred to as ''domestication selection' by Doyle (*Aquaculture*, Vol. 33, 1983).

Although salmonid culture has been practiced with some populations for a long time, there is only limited documentation of changes that have occurred as a result of the artificial propagation. Moreover, there is essentially no documentation of what factors may have contributed to changes that have occured; historical data concerning quantitative performance and the associated culture conditions are either not available or poorly recorded. This unavailablility of historical data is not unexpected because determining how to measure performance is one of the most difficult tasks in quantitative genetics. However, the absence of data has resulted in major confusion because the effects of the environment and the management system have been largely ignored. Instead of a careful examination of the total system of production, the general acceptance that most performance traits are under some degree of genetic control has often resulted in the unqualified conclusion that observed changes were caused by some form of intentional selection. As suggested in the title of this paper, one of our problems may be in understanding what we think we see!

Differences in performance among individuals in a population are said to be quantitative rather than qualitative because the differences are of degree rather than of kind. This qualitative nature of the differences means that performance characteristics show continuous variation from the extremes of lowest to highest performance. A portion of this observed variability is assumed to result from genetic differences among individuals but a major portion will be caused by dif-

[1] Aquaculture Program, Department of Animal Science, University of California, Davis, CA

ferential exposure of individuals to components of the environment. Using quantitative genetic terminology, the observed variability among phenotypes can be partitioned into two components, genetic variation and environmental variation.

The Environment

The concept of environmental effects is defined to include all those factors, other than the genotype of the individuals, which influence the performance of individuals in a population. Environmental effects are important in quantitative genetics because they mask genetic differences in a number of ways which in turn can result in improper or erroneous genetic interpretations. The most obvious effect of the environment is to influence average performance. It is generally recognized that if individuals of a single population are reared in two different environments, say cool versus cold water, there will be a real difference in the average growth of the two groups. Similarly, the average egg size of spawners often depends on the age of the females.

It is also possible for environmental differences to exist within what would otherwise be considered a single uniform management system. However, it does not follow that all individuals in the population will be randomly exposed to these environmental conditions. Sometimes it is possible to quantitate environmental effects according to definable categories while other effects are not definable and must be assumed to influence individuals randomly. The objective of quantitating as many environmental effects as possible is to provide a means of adjusting data on performance to remove the effects of non-random sources of environmental differences.

Salmonid producers recognize many of these sources of environmental differences in a qualitative way because they produce repeatable effects. Examples of environmental sources of variation which can potentially be quantitated include year of spawning, season of spawning, age of female, size of egg, date of birth, hatchery site, and method of rearing. At the present time there is little known about the quantitative effect of these sources of variability within a population.

The assumption is often made that the environment is uniform for all individuals or, in cases where individuals are known to have been treated differently, comparisons among fish are restricted to individuals treated equally. Making comparisons under such assumptions may not only result in incorrect decisions but also can eliminate comparisons that could demonstrate critical effects. For example, if it were known that age-2 and age-3 rainbow females in the same stock differed by 1000 in average number of eggs produced, it would be simple to correct the performance of all age-2 females to an age-3 equivalent for purposes of comparing individuals of different ages.

The ultimate danger of not understanding the nature of the environmental effects important in salmonid culture involves potential errors in defining traits and errors of interpretation of quantitative parameters. How and when a characteristic is measured constitutes an integral part of the definition of the trait and from a genetic point of view, the definition of the trait is also the definition of the genotype of interest. For example, if spawning date influences egg size, then comparing egg diameters for females spawned at different times includes a comparison of date of spawning as well as the genotypes of the females. Improperly defining a trait can be equivalent to measuring a characteristic other than the one intended which in turn can result in a lack of expected response to selection or in a response in an unexpected direction.

The Problem

This paper describes one example of what can be called a systematic environmental effect of spawning date on weight of rainbow trout at one year of age. In examining this effect we noted some unexpected relationships between reproductive traits which at first glance seemed to indicate unusual genetic differences. However, we conclude that there is an independent and direct effect of date of spawning on yearling weight.

In what follows, we will examine, in a subjective and qualitative way only, some interrelated observations on the reproductive performance of an age-2, fall spawning stock of rainbow trout. The data used for the discussion came from three consecutive, non-overlapping generations and consists of information on the spawning performance of dams and their daughters and the body size of offspring of the daughters. All ages were calculated as time, in days, from date of fertilization of the eggs.

To set the stage for the discussion, we will accept a commonly held notion that small eggs produce small fry that grow slower than fry from large eggs. This observation has led to the generalization that fish from small eggs must be genetically inferior to fish from large eggs. Carrying this logic one step further, and knowing that eggs obtained early in the spawning season are small relative to eggs obtained later in the season, one must conclude that early spawning females produce inferior offspring. The real question we need to address is, are the observed differences in average weight of fish at one year of age due to genetic differences or some environmental effect associated with date of spawning?

In an attempt to examine the effect of date of spawning (date eggs were fertilized) on weight of offspring. data on body weight at specific ages were sorted into weekly classes based on the date the daughter spawned. Similarly, data on the spawning characteristics of both the daughters and their dams were sorted according to daughter spawn date. Because of small numbers, the first spawning interval included all females spawning from the 1st through 26th of August; however, the majority spawned during the last 10 days of the interval. Each generation was identified by the year the females were spawned.

Body Size

In all three generations, the average 168-day weight of offspring was observed to increase as the spawning season progressed from August through October (Table 1). The average weight of fish from eggs spawned in October was over 40% greater than that of fish from August eggs. There were also major differences between the overall average weights for the three years. The comparison of years provides one example of how environmental effects may complicate the interpretation of results. For example, the average observed weight for 1980 may be low because spawning was completed by the end of September. The weight gain over the 28-day period from 168 to 196 days was also generally higher for the fish from later spawned eggs as was percent gain.

Similar differences in average weight were observed when the fish had

Table 1. Average weight of rainbow trout at 168 days (from date of fertilization) and absolute growth and percent gain over a 28-day period (168 to 196 days). The data are classified according to date of spawning of females in each of three years (generations).

SPAWN INTERVAL	WEIGHT (gm)			GROWTH (gm)			GAIN (%)		
	1978	1980	1982	1978	1980	1982	1978	1980	1982
8/1-8/26	2.54	1.97	2.82	3.14	2.27	3.20	123.6	115.2	113.5
8/27-9/2	—	2.42	2.83	—	2.76	2.97	—	114.1	104.9
9/3-9/9	2.07	2.55	3.05	6.10	2.66	4.00	294.7	104.3	131.1
9/10-9/16	—	2.51	2.77	—	3.43	3.79	—	136.6	136.8
9/17-9/23	2.80	2.80	3.15	3.35	4.19	5.38	119.6	149.6	170.8
9/24-9/30	—	3.04	3.69	—	4.59	5.30	—	151.0	143.6
10/1-10/7	3.56		3.69	4.14		7.78	116.3		205.4
10/8-10/14	4.22		—	4.43		—	105.0		—
10/15-10/21	4.19		—	4.43		—	97.4		—
10/22-10/28			4.66			9.25			198.5
AVERAGE	3.80	2.64	3.51	4.15	3.62	5.68			

Table 2. Average body weight of rainbow trout at 308 days of age and absolute growth and percent gain over a 28-day period (280 to 308 days). The data are classified according to date of spawning of females in each of three years (generations).

SPAWN INTERVAL	WEIGHT (gm)			GROWTH (gm)			GAIN (%)		
	1978	1980	1982	1978	1980	1982	1978	1980	1982
8/1-8/26	39.6	37.8	54.6	16.9	13.4	25.0	42.7	35.4	45.8
8/27-9/2	—	45.1	59.1	—	13.7	25.7	—	30.4	43.5
9/3-9/9	45.7	45.3	67.2	17.8	11.6	23.3	38.9	25.6	34.7
9/10-9/16	—	49.3	69.5	—	14.1	23.0	—	28.6	33.1
9/17-9/23	59.1	52.2	68.9	21.1	14.7	20.7	35.7	28.2	30.8
9/24-9/30	—	60.9	71.7	—	17.7	20.7	—	29.1	28.9
10/1-10/7	62.0		79.6	21.5		26.2	34.7		32.9
10/8-10/14	66.8		—	21.4		—	32.0		—
10/15-10/21	78.1		—	21.4		—	32.0		—
10/22-10/28			101.0			37.5			37.1
AVERAGE	66.5	50.3	75.0	22.1	14.5	24.8			

reached 308 days of age (Table 2). August spawned fish were almost 50% lighter than October fish. Differences among years were also evident but the ranking of the three years had changed from that observed at 168 days. The average weight for 1980 remained the lowest, however, the 1982 year, which was intermediate at 168 days, was the largest at 308 days. Differences in average gain in weight were less evident by 308 days resulting in the percent gain of earlier spawned fish being greater than that of the late fish. It should be noted that changes in obvious environmental parameters, such as changes in water temperature or density of fish, were not sufficient to explain the observed differences. Thus, if the effects were environmental, they would be best classified as "season effects."

Egg Size

Suspecting the effect may be related to egg size, we calculated the average egg size of the daughters spawning in each interval (Table 3). The first inconsistency evident from this analysis was the lack of real differences in the overall average egg size among the three years, suggesting that the differences in yearling weight are not easily explained by egg size. However, daughters that

Table 3. Average egg size (#/30 ml) for daughters sorted by date daughters spawned (Date Spawned) and by date daughters were conceived (Date Fertilized) for three consecutive generations of an age-2 spawning rainbow trout broodstock.

SPAWN INTERVAL	DATE SPAWNED			DATE FERTILIZED		
	78	80	82	78	80	82
8/1-8/26	519	410	457	359	384	439
8/27-9/2	—	362	431	374	—	398
9/3-9/9	432	419	422	379	359	386
9/10-9/16	—	356	415	—	—	389
9/17-9/23	357	365	386	358	355	397
9/24-9/30	—	350	365	348	—	355
10/1-10/7	362		367	—	359	
10/8-10/14	338		—	323	365	
10/15-10/21	344		—		373	
10/22-10/28			366			
AVERAGE	362	368	387	362	368	387

spawned earliest in each year did produce smaller eggs than those spawning later. The range in egg size was low in 1980 so this generation did not show a strong seasonal trend.

To examine the possibility that the trend in egg size over the spawning season was due primarily to genetic differences, we calculated the average egg size for daughters based on the date they were born (spawning intervals refer to date the daughter's dam spawned). This sorting produced averages representing the size of eggs produced by all daughters conceived during a particular interval. If the apparent seasonal trend was primarily due to genetic differences, then daughters out of early spawning dams should produce smaller eggs than daughters out of late spawning dams.

These averages are given under the heading of "Date Fertilized" in Table 3. There was a tendency for eggs produced by daughters out of early spawning dams to be smaller than those out of later spawning dams. However, the differences were minor in comparison to the trend observed for spawn date of daughters. This suggested that each group of daughters represented a nearly random genetic group regardless of their birth date. Therefore, genetic differences between groups of daughters could not be the major cause of the differences in egg size. The smaller differences associated with the time their dams spawned are probably expressions of genetic differences.

Sexual Maturity—Females

We then wonder why the early spawning daughters tended to produce smaller eggs. One possible explanation considered was the age at which daughters reach sexual maturity. The average age of daughters at spawning was calculated for each spawn interval (Date Spawned, Table 4). It was very evident that early spawning daughters were considerably younger than those spawning at the end of the season. The seasonal trend was large for 1978 and 1982 with a range in age at spawning of about 50 days. Variability was again low for 1980, an observation similar to that observed for egg size in this generation. The average age at spawning based on the date the daughters were born also showed a seasonal trend (Date Fertilized, Table 4). However, in this instance daughters out of early spawning dams spawned at an older age than daughters out of late

Table 4. Average age at spawning for daughters sorted by date daughters spawned (Date Spawned) and by date daughters were conceived (Date Fertilized) for three consecutive generations of an age-two spawning rainbow trout broodstock.

SPAWN INTERVAL	DATE SPAWNED			DATE FERTILIZED		
	78	80	82	78	80	82
8/1-8/26	723	714	719	777	751	753
8/27-9/2	—	719	738	765	—	750
9/3-9/9	723	705	733	765	736	743
9/10-9/16	—	709	735	—	—	744
9/17-9/23	751	715	739	758	716	747
9/24-9/30	—	721	743	748	—	735
10/1-10/7	762		747	—	709	
10/8-10/14	764		—	741	705	
10/15-10/21	772		—		696	
10/22-10/28			767			
AVERAGE	761	713	743	761	713	743

spawning dams. These results further support the idea that there is a compensation over generations for time of spawning and that there is a strong tendency for fish to spawn during the seasonal "peak." Stated another way there is a strong regression toward the mean.

It is also worth noting that in none of the three generations did the females spawn at an average of two years (730 days). The fish spawned in 1980 were extremely young, averaging only 713 days of age. This may possibly have been a compensation for the lateness of the spawning season of the previous generation (761 days). It is also possible the average age of spawning for 1982 daughters (743 days) was greater than 730 days because their dams spawned early. Average performances for the three generations also suggested that age at which the females spawned did not influence average egg size. Average egg size was very similar in 1978 and 1980 whereas the average age at sexual maturity was extremely different. Two conclusions seem appropriate from these analyses: (1) The differences observed in average age at spawning must have been due primarily to environmental effects and (2) it is possible to attribute all of the observed difference in egg size to age of the daughters.

Sexual Maturity—Males

Age of maturity of males has often been considered to effect size of fish. The generalization most frequently expressed is that precocious males are small and therefore must be genetically inferior in terms of their growth rate capabilities. We have been following the characteristics of a limited number of so-called "precocious males" in our rainbow trout broodstock that had been kept in the stock for reasons other than the age they reached sexual maturity. Males which were mature at one year of age were identified in 28 full sib families. Their average body weight was compared to the average weight of all full sibs in their respective families, including the mature males since it was not possible to sex the immature individuals.

Based on family means, the 59 males mature at one year were found to be almost 15% heavier than the average weight of all fish in the family (Table 5). When the stock reached two years of age, the 49 early maturing males still present were compared to the average weight of their brothers which were not mature at one year (referred to as "normal" brothers). Those males that had matured early were found to be slightly smaller than their full brothers although the difference was small.

It would appear that early maturity is associated with rapid growth rate and that the onset of sexual maturation results in a disruption of normal growth. This effect of sexual maturation on body weight can most appropriately be classified as an environmental effect. Our estimates of the effects of early maturity should be conservative. The full-sib family means for yearling weight included the early maturing males so these values may be larger than if the early maturing males had been excluded. Secondly, some of the males classified as "normal" two-year spawners may have matured prior to one year and therefore were not classified as precocious, an event which would affect the mean weight of the two-year-old "normal" brothers.

Table 5. Comparison of body weight of males maturing at one year of age with their sibs and with their "normal" two year maturing brothers based on full-sib family means.

	ONE YEAR AGE	TWO YEAR AGE
Number full-sib families	28	28
Number precocious males	59	49
Precocious males/family	2.1	1.8
Sibs (brothers/family)	30	6.1
Average weight precocious males (gm)	134.2 + 3.9	796.4 ± 30.3
Average weight sibs (brothers) (gm)	117.1 ± 3.1	826.8 ± 26.1
Difference (precocious - normal) (gm)	17.1	−30.4
Percent difference (%)	14.6	3.8

Table 6. Average 308-day weight of offspring for three generations (Table 2) adjusted for year effects. The entries represent average weight of fish according to the date daughters (mother) spawned (rows) and the date the daughter was conceived (columns). Blanks represent missing data.

SPAWN INTERVAL	8/1-8/26	8/27-9/2	9/3-9/9	9/10-9/16	9/17-9/23	9/14-9/30	10/1-10/7	10/8-10/14	10/15-10/21	ALL
8/1-8/26	57.2	56.8	59.8	47.5	53.7		41.9			52.1
8/27-9/2	65.9		50.2		58.1			61.8		59.0
9/3-9/9	73.2	61.6	71.4	63.4	65.0		59.6	63.0	61.5	64.8
9/10-9/16	66.3	66.9	63.6	71.4	67.8	75.5	64.9	68.7	65.7	67.9
9/17-9/23	66.0	73.8	77.4	69.1	75.1	69.2	74.9	71.1	69.4	71.8
9/24-9/30	70.1		51.8	74.1	79.2	71.7	81.7	84.7	75.9	73.6
10/1-10/7	84.6	83.2	80.0	80.0	77.6	82.2				81.3
10/8-10/14	89.6	85.2	99.2		90.2	84.6		89.1		89.6
10/15-10/21	95.1	96.3	103.8		110.1	106.4		110.5		103.7
10/22-10/28		97.6		104.3	98.1	104.3				101.1
AVERAGE	73.7	77.7	73.0	72.8	77.5	84.8	64.6	78.1	68.1	

Conclusions

As a final examination of the problem, the data on 308-day weight for the three generations were combined into one data set after being adjusted to remove the effects of years. Table 6 presents a summary of the data originally described in Table 1. The data were sorted simultaneously for both the date of spawning of the daughters that produced the fish and the date of spawning of their dams (date daughters were conceived).

Care must be taken in interpreting these results because of the large number of missing cells in the data matrix. However, some general observations can be made. There were large differences between the average weight of fish from daughters spawned in the early and late periods of the season (row averages, Table 6). On the other hand, there were few and only rather inconsistent differences when the data were classified by date of spawning of the dams (column averages, Table 6). Therefore, it seems safe to conclude that the seasonal trend in weight was not due to differences in the genetic value of the daughters spawning at different times. If there were genetic differences among the daughters, a similar although smaller seasonal trend should have been evident from data classified by spawn time of dams.

Thus, we can see that "things are not always what they seem to be." Having come full circle, we still have no clear explanation for early fall spawning fish being smaller than those from eggs obtained later in the spawned season. With these examples, an attempt has been made to show how interpretation of genetic data can be complicated by poorly understood relationships among various traits and between environmental and genetic effects. Only qualitative results and interpretations have been suggested. A complete assessment will be dependent on rigorous quantitative analyses of large data sets obtained from diverse broodstock, similar to that presented by Gjedrem (these proceedings). However, it is quite evident that there is a need for these studies if the application of quantitative genetics to the improvement of salmonid reproduction is to become completely effective.

GENETIC VARIATION IN AGE AT MATURITY AND ITS RELATION TO GROWTH RATE

Trygve Gjedrem[1]

Abstract: Age at sexual maturity is of great economic importance in sea ranching and in fish farming. In salmonids there is considerable variation between species as well as between strains within species. Therefore problems can be partly overcome by choosing stocks suitable for a particular production system.

It is remarkable that the number of years in fresh water, ranging from 1 to 7 depending on water temperature or latitude, seems to have very little influence on the time in sea before maturation takes place in Atlantic salmon.

Genetic variation in age at maturity is quite high with a heritability of around 0.3. This means that it is possible to change this trait by selection.

Maturation of parr is also a heritable trait. However, it seems to be independently inherited from age at maturity.

There seems to be negative phenotypic as well as genetic correlation between age at maturity and body weight. This indicates that when selecting for increased growth rate the age at maturity may be reduced. In fish farming, when producing a large fish, this is not desirable. Combined selection for both traits simultaneously should therefore be practiced.

Introduction

In the farming as well as in sea ranching of salmonids, age at maturation is of considerable economic importance. Since flesh quality deteriorates, the fish must be marketed before reaching maturation. Also, size of fish when slaughtered is partly dependent on age at maturation. Size also affects the price of the product. Consequently, there is great interest in developing late maturing populations for farming, and in addition, geneticists are actively interested in producing sterile fish to forego these maturation-related problems.

In sea ranching of salmon it is not obvious whether early or late maturing strains are preferable. According to Hansen (1983 unpublished) Norwegian grilse rivers have a much higher return frequency than rivers with late maturing strains of Atlantic salmon. Saunders (1979) proposed use of grilse strains to avoid losses to distant commercial fisheries. Thus there seems to be an inverse relation between size at capture and return frequency in sea ranching because early maturation means small fish (1-3 kg) and late maturing strains yield large salmon.

It is therefore of considerable interest to discuss what possibilities there are to change age at maturation in salmonid populations to produce a more pro-

[1] Department of Animal Genetics and Breeding, Agricultural University of Norway, As-NLH

ductive fish for farming and sea ranching. In this connection the magnitude of genetic variation is quite decisive. Because body size of fish is also of such great economic importance the genetic correlation between age at maturation and growth rate is also of considerable interest.

Age at Maturation

In domestic animals size is one of the major factors for determining age at maturation although it can be modified by environmental factors which can create quite large variations within species. In anadromous salmonids the freshwater life histories vary considerably from one species to another. Some of the Pacific salmon (pink and chum) migrate almost immediately into brackish or sea water, while some of the Atlantic salmon strains from rivers with low temperatures may reside up to 7 years in fresh water before migration.

Hutton (1937) working with Atlantic salmon proposed what he called "the inverse relation theory of river and sea-life." This postulates that the total age at maturity tends to be constant. However, results from three Norwegian rivers (Dahl, 1937), five North American (Ritter, 1974) and fourteen Scottish, Welsh, and Irish (Gardner, 1976) show, on average, that the smolt age of fish returning as grilse (1 year at sea) are 0.07 years older than salmon returning after 2 years at sea and 0.13 years older than salmon returning after 3 years at sea (Table 1). There is considerable variation between rivers and areas both in age of smolt as well as time at sea. These results lead to the conclusion that the age at maturation of Atlantic salmon is not determined by the number of winters in fresh water.

Supporting Hutton's theory are several reports (Ritter, 1975, Ritter and Newbould, 1977, Saunders et al., 1981 and Hansen and Lea, 1982) of farmed smolts which show significantly lower grilse frequency in 1-year-old smolts compared to 2-year-old smolts.

Ricker (1972) in discussing possible sources of variation of length of sea life concluded that there is a trend for all species of *Oncorhynchus* toward longer time in the ocean with increasing latitude.

Gardner (1976) reviewed the literature on differences among sexes in age at maturation and concluded that the tendency for males to mature earlier than females apparently applies to all salmonids.

There is considerable variation in age at maturation in salmonid fishes. It is well known that many of the Atlantic salmon strains have some precocious males in fresh water even as underyearlings with bodyweights as low as 10-15 g. It has also been reported (Rutter, 1904) that chinook males mature as underyearlings.

Table 1. Average smolt-age of Atlantic salmon returning to freshwater after different number of years at sea. (Reported by Gardner 1976)

AREA	NUMBER OF RIVERS	WINTERS AT SEA 1	2	3
Norway	3	3.21	3.26	3.37
North America	5	3.21	3.04	2.90
Scotland, Wales, Ireland	14	2.20	2.14	2.08
Average	27	2.57	2.50	2.44

The phenotypic variation in age at maturation in the sea is considerable. Hansen (1983, unpublished) reports the incidence of maturation in Atlantic salmon after one summer in the sea. Gjerde (1984), using 1year-old smolts for cage culture of Atlantic salmon, reports variation from 1-1/2 to 5-1/2 sea years. According to Ricker (1982) the Pacific salmon attain sexual maturity at the following ages including fresh water life: Pink 2 years, chum 3-5 years, sockeye 3-6 years, coho 2-3 years and chinook 2-6.

It is reasonable to believe that environmental sources of variation are responsible for this large phenotypic variation in age at maturation. Besides water temperatures, both in fresh and sea water, availability of food is likely to be of particular importance.

Genetic Variation in Age at Maturation

Precocious males: Several investigations have found differences in frequency of precocious males between salmonid strains (Naevdal et al., 1978 and Saunders and Sreedharan, 1978). Ricker (1972) concludes that genetic differences between strains in presence or absence of precociously mature parr exist. Naevdal (1983) found significant variation between families of Atlantic salmon in frequencies of precocious males. Thorpe et al. (1983) found that incidence of maturity as parr is a heritable trait.

Gjerde (1984), in a selection program with Atlantic salmon used milt from parr, 4- and 5-year-old males and fertilized eggs from 4- and 5-year-old females. Progenies of precocious males had higher frequencies of precocious males compared to ordinary males. Gjerde (1984) also found evidence that maturation of precocious males and age at maturation in the sea are two independently heritable traits. This is in agreement with findings of Glebe et al. (1980).

The magnitude of genetic variation or heritability for frequency of precocious males is not known. However, there seems to be no doubt that this is a heritable trait. This causes some concern in fish farming, and is said to be of importance for maintaining wild strains.

Strain variation: Ricker (1972) reviewed the literature and concluded that genetic factors play a major part in the determination of age at maturation of adult Pacific salmon. In Atlantic salmon it is evident that there are large differences among river strains in age at maturation. This is clearly shown in Figure 1 by Gunnes (1978).

Some rivers are typical grilse rivers with the majority of fish maturing after one winter in sea, while others are salmon rivers in which fish mature at older ages. Some of this variation between strains can be environmental because of differences in water flow and water quality. However, it is likely that such environmental differences among river systems has been the base for natural selection which has resulted in genetic differences between strains.

In several experiments different salmon strains, known to differ naturally for age at maturation, have been reared under standardized farming conditions during their whole life. The results showed that there is significant variation among strains in age at maturation even in the absence of environmental differences (Gunnes, 1978, Naevdal et al., 1978, Gjerde and Gjedrem, 1983; Gjerde and Refstie, 1983). This clearly indicates that there is some genetic variation among strains for age at maturation. However, Gjerde and Gjedrem (1983) esti-

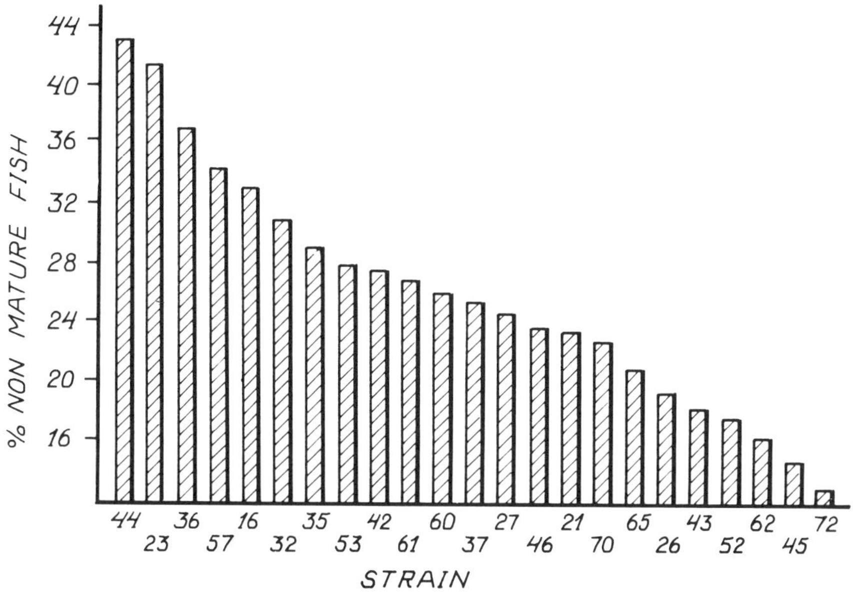

Figure 1. Variation in frequency of maturation of different Atlantic salmon strains after two sea winters (Gunnes, 1979).

mated the variation due to strain differences was only 1.5% of the total phenotypic variation.

Family variation: In planning a selection program for each trait it is of great importance to know the proportion of total phenotypic variance that is due to genetic variation; this is termed 'heritability'. Until now the most extensive study on age at maturation has been conducted by Gjerde and Gjedrem (1983). The data are comprised of information from 83 sires and 217 dams of Atlantic salmon and 56 sires and 108 dams of rainbow trout. In Atlantic salmon 12 strains were involved and the genetic analysis was done on a within strain basis. As shown in Table 2, the heritability based on sire component is high for Atlantic salmon and of moderate size for rainbow trout.

Busack and Gall (1983), working with fullsib groups of mosquitofish, estimated the heritability for age at maturation to be of the same level as that found for Atlantic salmon. Ricker (1980a) estimated heritability for age at maturation in chinook to be 0.30 based on data from L. Donaldson and his associates.

Naevdal et al. (1978) and Naevdal (1983) found significant differences between families of Atlantic salmon in age at maturation; Similar results were obtained for families of rainbow trout by Naevdal et al. (1981) and Naevdal (1983).

This means that, with the fairly large phenotypic variation and the heritability estimates, there are good possibilities to change age at maturation in the desired direction by means of selection, particularly for Atlantic salmon. Thus, the opportunity is present for changing this trait in fish farming or sea ranching operations.

The few estimates of heritability indicate that those based on the dam component are a little higher than those based on sire component. Therefore,

Table 2. Estimates of heritability for age at maturation.

SPECIES	HERIT-ABILITY	METHOD OF ESTIMATION	AUTHOR
Atlantic salmon	0.39	sire component	Gjerde and Gjedrem (1983)
” ”	0.49	dam component	” ” ” ”
” ”	0.48	regr. prog./dam	Gjerde (1984)
Rainbow trout	0.21	sire component	Gjerde and Gjedrem (1983)
” ”	0.26	dam component	” ” ” ”
Mosquitofish	0.41	fullsib component	Busack and Gall (1983)

some nonadditive genetic variance may be present. However, if the values given in Table 2 are true estimates of heritability, the differences between sire and dam components are so small that it should not justify large efforts to try to utilize the nonadditive genetic variance in breeding programs.

Selection: Selection programs and selection experiments give the most reliable estimates of the magnitude of genetic variance. Gjerde (1984) selected parents according to age of individual at maturation. Some of the results are presented in Figure 2. The frequency of early maturation was higher in families from young parents (4 x 4) compared with those from older parents (5 x 5). The crosses between the two age groups gave intermediate results. The regression of age of progenies on age of dam (within sire) gave an estimate of heritability of h^2 = 0.48 (Table 2) which is in good agreement with Gjerde and Gjedrem (1983).

Donaldson and Olson (1957) reported a decrease in age at first spawning in a selection program with rainbow trout. Ellis and Nobel (1961) fertilized eggs of chinook salmon with milt of grilse (1 winter at sea) and another group with milt of salmon (2 winters at sea). The grilse male parent produced 16% grilse while salmon produced 7% grilse spawners. Elson (1973) using hatchery reared smolts of Atlantic salmon, found that grilse parents produced more grilse (61%) than salmon parents (39%). Piggins (1974) got the following results in a similar experiment; grilse parents gave 2.5% progenies that had been in the sea 2 years compared with salmon parents giving 20%. Ritter and Newbould (1977) found higher frequencies of grilse among progenies of grilse parents compared with progenies of salmon parents.

Ricker (1980a,b) studied changes in body weight and age at maturation for Pacific salmon populations based on records taken at fishing. For chum, Ricker (1980a) found an increase in age of 0.02 years per year or 0.8%. The body weight decreased by 0.03 kg per year or 0.6% over a 16-year period. For chinook salmon Ricker (1980b) found a different situation. Age at maturation was reduced by 1.5–2.0 years or 0.04% per year over a 50 year period. During this period the body weight of chinook was about halved or it was decreased by 0.1 kg per year.

According to Ricker the main reason for this reduction in body weight, which also took place for the other Pacific species, is the selective fishing practiced. Thus the selection "criterion" was body weight and the change in age at maturation was a correlated effect. For chinook this implies a positive and for chum a negative genetic correlation between body weight and age at maturity.

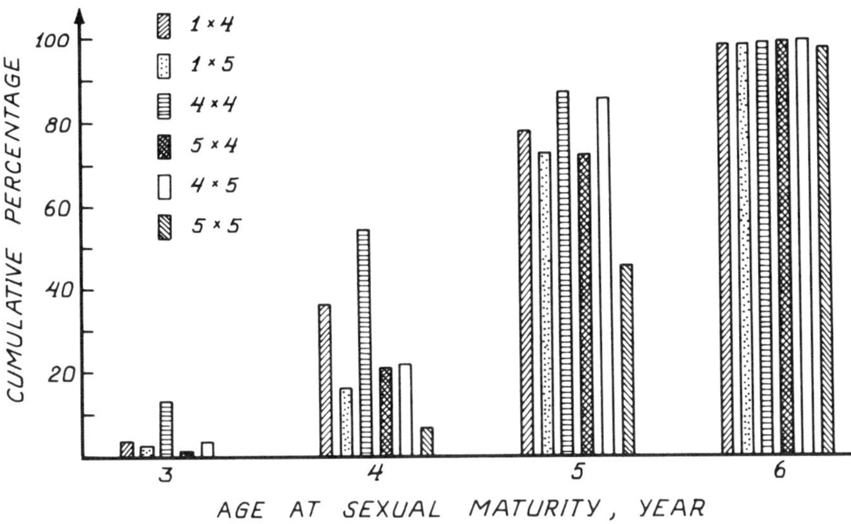

Figure 2. Each group cumulation percentage of sexual mature fish at 3, 4, 5 and 6 years of age (Gjerde, 1984).

Relationships Between Age at Maturation and Growth Rate

Phenotypic correlation: Ritter (1972) found that small smolts of Atlantic salmon gave higher grilse percentage than large smolts, which suggested a positive correlation between age at maturation and smolt size. This was the case for hatchery reared smolts while there was no clear correlation for wild smolts.

Hallingstad (1978) working with rainbow trout, reported high frequency of early maturation for fast growing fish at an age of 6–8 months compared with slow growing fingerlings which matured at older ages. This indicates a negative correlation between early growth rate and age at sexual maturation which is in agreement with Bilton (1980) working with coho and contrary to the result of Ritter (1972).

Alm (1959) concluded that the fastest growing fish in a population mature earliest. This is in agreement with Simpson and Thorpe (1976), Kato (1980) and Thorpe et al. (1983). The result obtained by Gjerde and Refstie (1983) and Gjerde (1984) showed that maturing fish were heavier than immature fish of Atlantic salmon, which means that there is a negative phenotypic correlation between these two traits. Naevdal et al. (1981) also reported that maturing fish on average were heavier than immature fish.

Gjerde and Gjedrem (1983) found a significant negative correlation between body weight and age at maturity in Atlantic salmon and a low negative correlation in rainbow trout (Table 3).

According to these results there seems to be a low negative phenotypic correlation between growth rate and age at maturation.

Genetic correlation: In a selection program it is the genetic correlation which gives correlated response in other traits and therefore it is important to have esti-

Table 3. Relationships between age at maturation and body weight 4–5 months prior to maturation (Gjerde and Gjedrem, 1983).

	PHENOTYPIC CORRELATION	GENETIC CORRELATION
Atlantic salmon	−0.32	−0.52
Rainbow trout	−0.11	−0.11

mates of genetic correlation between the traits under selection as well as with other traits of importance.

As already mentioned, Ricker (1980a,b) found correlated response in age at maturation in chum and chinook salmon resulting from selective fishing on size. However, the results in chum indicate a negative genetic correlation while that in chinook was positive. The size of the correlations were not estimated.

The only genetic correlation estimates available to date (Gjerde and Gjedrem, 1983) are shown in Table 3. They studied the frequency of mature fish after 2 years in the sea for Atlantic salmon and after 1–1¼ years for rainbow trout. The genetic correlation between age of maturation and body weight in Atlantic salmon is quite high, and rather low for rainbow trout; both are negative. It should be stressed that the body weight used was taken 4–5 months prior to spawning, which means that the development of gonads had already started.

Gjerde (unpublished), working with 132 fullsib families of Atlantic salmon, calculated correlations between family averages. Each family contained about 70 fish. The frequency of grilse was 31%, and of salmon, or fish maturing after two sea winters, was 72%, while average body weight was 4.5 kg. The correlations estimated were not true genetic correlations because the number in each family was limited. The results, which are given in Table 4, show a low and not significant correlation between early and late maturation. Further, there is a low positive significant correlation between early maturation (percent grilse) and body weight after two years in sea. However, the correlation between body weight and late maturation (percent salmon) is negative and of the same size as the genetic correlation given by Gjerde and Gjedrem (1983). Similar results were obtained using halfsib averages.

Discussion

In a selection program age at maturity should be included because of its economic importance. In a production system where large fish is the goal, selection for late maturing fish should be practiced. However, this will lengthen the generation interval and thus reduce the genetic gain per year. In such a situation production of sterile fish would be particularly useful.

Glebe et al. (1980) and Naevdal (1983) proposed use of precocious parr to accelerate genetic gain by shortening the generation interval. The difficulty which the authors did not discuss was how to estimate breeding values for economic traits in precocious males.

From the review of the literature it can be concluded that the phenotypic variation in age at maturation is quite large both between species and between and within strains. The few estimates of heritability available (Table 2) show high values for Atlantic salmon while those for rainbow trout are lower. Thus, the additive genetic variance is considerable. This is in agreement with the response obtained in selection experiments (Donaldson and Olson, 1957; Ellis and Nobel, 1961; Elson, 1973; Piggins, 1974; Ritter and Newbould, 1977; and Gjerde, 1984). Therefore, it can be concluded that the possiblities to change age at maturation through selection is good. Since age at maturation is an all or none trait, family selection will be much more efficient than individual selection. One should keep in mind that frequency of precocious males and age at later maturation seems to be independent (Gjerde, 1984) and that there seems to be a low correlation between early and late maturation (Table 4).

The estimates of heritabilities (Table 2) based on sire and dam components show only small differences, which indicate low nonadditive genetic variance in age at maturation. It is therefore not likely that breeding programs will gain much in efficiency by including crossbreeding.

According to the results presented, there seems to a quite strong negative genetic correlation (r_G = -0.5) between body weight and age at maturation (Tables 3 and 4). This is supported by the results of Ricker (1980a). However, one should take into consideration that body weight is recorded 4-5 months prior to spawning (Gjerde and Gjedrem, 1983 and Gjerde, 1983, unpublished). This will impose an automatic correlation between these two traits, which will yield a higher value for the phenotypic as well as for the genetic correlation. It is interesting to note that Gjerde (1983, unpublished) estimated a low positive correlation between grilse frequency and body weight of families one year later. The negative genetic correlation obtained could partly be influenced by the effect of sex hormones stimulating growth in maturing fish.

At present, it is not quite clear what are the causes and effects in the relationships between growth rate and age at maturation. As pointed out by Gjerde and Refstie (1983) it should be investigated whether this difference in body weight between mature and immature fish is caused by differences in growth ability *per se*, or whether the maturation process accelerates growth.

Table 4. Coefficient of correlation between age at maturation and body weight of Atlantic salmon (132 fullsib families) and rainbow trout (85 fullsib families). (Gjerde 1983 unpublished).

CHARACTER:	ATLANTIC SALMON	RAINBOW TROUT
% early m. and % late m.	0.12	0.21
% early m. and body weight	0.27	0.30
% late m. and body weight	−0.52	−0.27

1) **% early m.** = maturation after sea winter for Atlantic salmon and during the first sea winter for rainbow trout.

2) **% late m.** = maturation after 2 sea winters for Atlantic salmon and during the second sea winter for rainbow trout.

References

Alm, G., 1959. Connection between maturity, size and age in fishes. Rep. Inst. Freshwater Res. Drottningholm 40:5-145.

Bilton, H.T., 1980. Experimental releases of coho salmon in British Columbia. In: J.E. Thorpe (Editor), Salmon Ranching, Academic Press, London, pp. 305-324.

Busack, C.A. and Gall, G.A.E., 1983. An initial description of the quantitative genetics of growth and reproduction in the mosquitofish, *Gambusia affinis*. Aquaculture 32:123-140.

Dahl, K., 1937. Parr life of salmon. Salm. Trout Mag. 86:10-13.

Donaldson, L.R. and Olson, P.R., 1957. Development of rainbow trout broodstocks by selective breeding. Trans. Am. Fish. Soc. 85:93-101.

Ellis, C.H. and Noble, R.E., 1961. Return of marked fall chinook salmon to the Deschutes River, Deschutes River genetics experiment. Wash. State Fisheries Dept. Ann. Rept. for 1960, 70:72-75.

Elson, P.F., 1973. Genetic polymorphism in Northwest Mariamichi salmon, in relation to season of river ascent and age at maturation and its implications for management of the stocks. ICNAF Res. Doc. 73/76, 6 pp.

Gardner, M.L.G., 1976. A review of factors which may influence the sea-age and maturation of Atlantic salmon *Salmo salar*. L.J. Fish. Biol. 9:289-327.

Gjerde, B., 1984. Response to individual selection for age at sexual maturity in Atlantic salmon. Aquaculture, in press.

Gjerde, B. and Gjedrem, T., 1983. Estimates of phenotypic and genetic parameters for some carcass traits in Atlantic salmon and rainbow trout. Aquaculture, in press.

Gjerde, B. and Refstie, T., 1983. Complete dialell cross between five strains of Atlantic salmon. Livest. Prod. Sci., in press.

Glebe, B.D., Eddy, W. and Saunders, R.L., 1980. The influence of parental age at maturity and rearing practice on precocious maturation of hatchery-reared Atlantic salmon parr. ICES C.M. 1981/F:8, 8 pp.

Gunnes, K., 1978. Genetic variation in production traits of Atlantic salmon. EAAP-Commission on animal genetics. Stockholm, 1978, 7 pp.

Hallingstad, F., 1978. Grading and early maturation in rainbow trout (Sortering og kjönnsmodning hos regnbueörret). Nor. Fiskeoppdrett 3(1):9-10. (In Norwegian).

Hansen, L.P. and Lea, T.B., 1982. Tagging and release of Atlantic salmon smolts (*Salmo salar* L.) in the river Rana, Northern Norway. Rep. Inst. Freshwater Res., Drottningholm, 60:31-38.

Hutton, J.A., 1937. Wye parr and smolts: the inverse ratio theory of river and sea-life. Salm. Trout Mag. 87:119-123.

Kato, T., 1980. Relation between age at maturity and growth of kokanee, *Oncorhynchus nerka*, in Lake Towada. Bull. Nat. Res. Inst. Aquaculture 1:7-19.

Kazakov, R.V., 1981. Relation between sizes of adult Atlantic salmon (*Salmo salar* L.) and duration of river and sea-life periode. Aquaculture 24:327-340.

Naevdal, G., 1983. Genetic factors in connection with age at maturation. Aquaculture 33:97-106.

Naevdal, G., Holm, M., Ingebrigtsen, O. and Møller, D., 1978. Variation in age at first spawning in Atlantic salmon (*Salmo salar*). J. Fish. Res. Board. Can. 35:145-147.

Naevdal, G., Lerøy, R. and Møller, D., 1981. Variation in growth rate and age at first maturation in rainbow trout. Fisk. dir. Skr. Ser. Hav Unders. 17:71-78.

Piggins, D.J., 1974. The results of selective breeding from known grilse and salmon parents. Ann. Rep. Salm. Res. Trust Ireland XVIII, 35-39.

Ricker, W.E., 1972. Hereditary and environmental factors affecting certain salmonid populations, p. 27-160. In the stock concept in Pacific salmon. H.R. MacMillan Lectures in Fisheries, University of British Columbia, Vancouver, B.C.

Ricker, W.E., 1980a. Causes of the decrease in age and size of chinook salmon (*Oncorhynchus tshawytscha*). Can. Tech. Rep. Fish. and Aquatic Sci. No. 994; 25 pp.

Ricker, W.E., 1980b. Changes in the age and size of chum salmon (*Oncorhynchus keta*). Can. Tech. Rep. Fish. and Aquatic Sci. No. 930, 99 pp.

Ricker, W.E., 1981. Changes in the average size and average age of Pacific salmon. Can. J. Fish. Aquat. Sci. 38:1636-1656.

Ritter, J.A., 1972. Preliminary observation on the influence of smolt size on tag return rate and age at first maturity of Atlantic salmon (*Salmo salar*). ICES C.M. 1972/M:14, 9 pp.

Ritter, J.A., 1974. Relationships of smolt size and age at first maturity in Atlantic salmon. ICES C.M. 1974/M:25, 6 pp.

Ritter, J.A., 1975. Relationships of smolt size and age with age at first maturity in Atlantic salmon. Rev. Branch. Environ. Can. Tech. Rep. Ser. No. Mar/F-75.5, 7 pp.

Ritter, J.A. and Newbould, K., 1977. Relationships of parentage and smolt age to age at first maturity of Atlantic salmon (*Salmo salar*). ICES C.M. 1977/M:32, 5 pp.

Rutter, C., 1904. Natural history of the quinnat salmon. A report of investigations in the Sacramento River, 1886-1901. Bull. U.S. Fish Comm. for 1902, p. 65-141.

Saunders, R.L., 1979. Sea ranching—A promising way to enhance populations of Atlantic salmon for angling and commercial fisheries. NASRC Report No. 8:8 pp.

Saunders, R.L., Henderson, E.B. and Glebe, B.D., 1981. Evidence of a major environmental component in determination of the grilse: large salmon ratio. ICES C.M. 1981/M:29, 12 pp.

Saunders, R.L. and Sreedharan, A., 1978. The incidence and genetic implications of sexual maturity in male Atlantic salmon parr. NASRC Report No. 2: 8 pp.

Schaffer, W.M. and Elson, P.F., 1975. The adaptive significance of variations in life history among local populations of Atlantic salmon in North America. Ecol. 56:577-590.

Simpson, T.H. and Thorpe, J.E., 1976. Growth biomodality in the Atlantic salmon. ICES C.M. 1976/M:22, 7 pp.

Thorpe, J.E., Morgan, R.I.G., Talbot, C. and Miles, M.S., 1983. Inheritance of developmental rates in Atlantic salmon, *Salmo salar* L. Aquaculture 33:119-129.

ABSTRACTS–Genetics

The Influence of Genetic and Environmental Factors on Maturity in Farmed Atlantic Salmon

R. Alderson, Marine Harvest Ltd., Lochailort, Inverness-shire, United Kingdom

The physiological changes associated with maturation in salmon can place a severe constraint on the farmer's options for selling his fish. If they are not to be used for breeding, the farmer must sell them before there is any deterioration in appearance and flesh quality. Problems can occur; if too many fish mature and cause a glut, or if the fish are too small for the market's requirements. These problems can be particularly acute with fish maturing as grilse only 12–14 months following transfer to sea.

In choosing and developing his stock the farmer therefore needs to understand to what extent genetic and environmental factors control the proportion of his fish which will mature each year. Only then can he predict, and eventually control, the pattern of harvest he will need to follow.

Marine Harvest Ltd., who have been farming Atlantic salmon in Scotland since the late 1960s, began a detailed program in 1977 to compare the performance of various stocks of both farmed and wild origin. Each stock, taken as eggs, is reared in replicated tanks up to the smolt stage when fish are cold branded, mixed, and transferred to seawater cages. Up to 13 stocks are tested each year and aspects of performance, including the incidence of maturity in the first and second years at sea, are noted.

This program has given a considerable body of information on the patterns of maturity shown by fish of different origins under farmed conditions. The results to be presented will show the relative importance of genetic and environmental factors in determining the pattern of maturity within the range of sites occupied by Marine Harvest on the west coast of Scotland.

Genotypic and Environmental Effects on the Incidence of Sexual Precocity in Coho Salmon

R. N. Iwamoto, B. A. Alexander, and W. K. Hershberger, University of Washington, Seattle

The relative effects of genotype and initial freshwater rearing on the incidence of sexual precocity in coho salmon males were examined. Eggs from normal females were fertilized with milt from both normal and precociously maturing males (jacks) in a partially nested factorial mating design. Progeny from each of the resulting full- and half-sib families were subsequently reared under two rearing temperature treatments (ambient and 15° C) during part of the hatchery phase of the life cycle.

Mean growth of progeny from the jack and normal-sired groups was equivalent during the majority of the initial freshwater rearing period, although a slight but significant increase in size of jack-sired progeny was detected at the end of the accelerated hatchery rearing period. Of the 4980 survivors at the termination of the experiment, 175 individuals or 3.75 percent of the population had precociously matured. These jacks were distributed disproportionately among jack-sired and normal-sired families with jack-sired families contributing a significantly higher number (4.6 times as many) of jacks. Despite significant temperature effects in terms of initial freshwater growth, the subsequent incidence of jacks did not appear to be temperature related. These results and the genetic analysis of the relative importance of genotype (category of male parents and individual paternal and maternal effects) on the incidence of sexual precocity are presented.

Heritability and Early Expression of Ovarian Recruitment Processes in Different Forms of Atlantic Salmon *(Salmo salar)*

A. M. Sutterlin and D. A. MacLean, Memorial University of Newfoundland, St. John's, Canada

The two pure lines and the reciprocal hybrids between an early maturing, dwarf form of landlocked Atlantic salmon and a later maturing anadromous form were reared in captivity for a minimum period of 2 years. Ninety percent of female parr of the dwarf landlocked salmon matured at age 2-plus years (fork length 15 cm), while no females of the anadromous form are expected to mature at this time. The size of the ovary differed in the two pure forms at age 0-plus, and the numbers, size and stages of previtellogenic oocytes also differed at age 1-plus. Ovarian patterns in the two hybrid forms more closely resembled that of the maternal parent. The implications of a genetically controlled predestinate influencing rates of sexual maturation, smoltification and subsequent fecundity are discussed.

A Conceptual Fitness Model for Managing Pacific Salmon Fisheries: Description of the Model

A. R. D. Kapuscinski and J. E. Lannan, Oregon State University, Newport

As information relating to the population genetics of Pacific salmon becomes more and more comprehensive, it is increasingly apparent that a conceptual framework must be developed which permits the translation of this information into management principles. This report describes a conceptual model for managing Pacific salmon fisheries to maintain a long term reproductive fitness in breeding populations. The model facilitates estimating the requirement for escapement to maintain reproductive fitness at a predetermined level, or determining if a given escapement results in a change of population fitness.

The model may be adjusted to accomodate the different life histories and reproductive strategies of the several species of Pacific salmon, and it considers both the mean and variance of fitness.

A Conceptual Fitness Model for Managing Pacific Salmon Fisheries: Application of the Model

J. E. Lannan and A. R. D. Kapuscinski, Oregon State University, Newport

Application of the model presented in the previous paper to the management of hypothetical salmon populations is demonstrated by simulation. The demonstration includes the estimation of required escapement to maintain reproductive fitness and the preselected levels of escapement which will result in changes in reproductive fitness of a hypothetical population. Algorithms employed in the model are briefly reviewed.

Production, Survival, and Growth of Triploid Trout

G. H. Thorgaard, Washington State University, Pullman

There has been considerable interest recently in induced triploidy in trout because sterile triploids might show better growth and survival at maturity than normal fish, and because of the potential for control of reproduction. Triploidy can be induced by heat shock in several trout species. Heat shock techniques can be readily adapted to large scale production. Triploids within a species show slightly lower survival than diploids, while triploid hybrids show better survival than diploid hybrids. Triploid rainbow trout show evidence of better growth than normal fish after sexual maturity.

Chromosome Set Manipulation in Three Species of Salmonids Using Hydrostatic Pressure

S. K. Allen, Jr., and J. M. Myers, University of Washington, Seattle

The use of hydrostatic pressure to affect early developmental events has been examined in various taxa, but few investigations have dealt with the potential practical utility of this method. Intensity and duration of hydrostatic pressure treatments were varied to estimate the optimal parameters for production of triploid and gynogenetic salmonids. These parameters were then applied to 5 families in 3 different species: chinook *(Oncorhynchus tshawytscha)* and coho salmon *(O. kisutch)* and rainbow trout *(Salmo gairdneri)*. Pressures of 10,000 psi for 8–10 minutes yielded the best compromise between conversion and mortality. Conversion rate across families was more consistent than parallel heat treatments and generally exceeded 90% in all groups. Mortality was more variable in pressure treatments than heat treatments. At present, pressure treatments offer an alternative to heat for experimental chromosome set manipulation, especially where consistency among replicates is desired. Prospects for the application of pressure to production schemes are discussed.

Production of Triploid Landlocked Atlantic Salmon *(Salmo salar* L.*)* and the Implications of Their Haemotology to Oxygen Utilization

T. J. Benfey and A. M. Sutterlin, Memorial University of Newfoundland, St. John's, Canada

Triploidy was induced in landlocked Atlantic salmon with 100% success and 70–90% survival (relative to controls) using either heat shocks or hydrostatic pressure. Triploidy was rapidly and accurately assayed at either the alevin or parr stage using a channelized Coulter Counter to measure the volume of a minimum of 150,000 erythrocytes per fish from a 1 μl sample of whole blood. Triploids had a larger mean erythrocyte volume (MCV) but lower erythrocyte count than diploids, and haematocrit was the same for both diploids and triploids. The haemoglobin content of the triploid blood was lower than that of the diploids, however, mean corpuscular haemoglobin (MCH) was higher due to the larger MCV. Triploids had a lower mean corpuscular haemoglobin concentration (MCHC), however, this was complicated by the fact that the nucleus of triploid erythrocytes occupied a greater percentage of corpuscular volume than did the diploid nucleus. MCHC was found to be the same for diploids and triploids when recalculated on a cytoplasmic rather then corpuscular level. The increase in MCV of triploids was due mainly to an increase in cell length; there was only a minor increase in cell width and no increase in cell height. Triploids had a lower rate of oxygen consumption than diploids; however, there was no difference in the oxygen tension at asphyxiation.

Artificial Polyploidization of Salmonids by Hydrostatic Pressure

Hiroshi Onozato, Hokkaido University, Hakodate, Japan

Artificial polyploidization technique is important to produce parthenogenetic diploids and polyploids. Thermal shock is most widely used to duplicate the chromosome set and it is very successful in some species. In salmonids, however, many eggs subjected to the shock remain unaffected.

In the present study, it is found that hydrostatic pressure has a marked effect on polyploidization of salmonid eggs. Gynogenetic diploids, androgenetic diploids, triploids and tetraploids were successfully produced in rainbow trout, masu salmon and chum salmon using hydrostatic pressure.

Optimum amount of pressure and the duration of treatment were tested on rainbow trout eggs inseminated with genetically inactivated sperms by UV irradiation. Eggs were treated 5 min. after insemination. 100% of eggs were diploidized when eggs were subjected to hydrostatic pressure (more than 650 kg/cm^2) for more than 6 min. Although all embryos treated were diploid, 36.9 to 58.9% of embryos were abnormal in that their body shape was similar to haploid embryos. Optimum time after insemination at which eggs were subjected to pressure shock (650 kg/cm^2, for 6 min.) was determined in chum salmon and masu salmon. Eggs were inseminated with genetically inactivated sperms and maintained at 10° C until treatment. Normal diploid embryos were encountered when eggs were treated during 5 to 40 or 270 to 330 min. after insemination. The ratios of normal embryos to developing ones were 58.3 to 78.6% and 16.7 to 40.0%, respectively.

When rainbow trout eggs inseminated with normal sperms were subjected to hydrostatic pressure (700 kg/cm^2, for 7 min.) at 15 min. after insemination, 100% of embryos were triploid. The ratio of normal embryos at eyed stage varied from 58.6 to 83.0%. Tetraploids were induced when eggs were treated (700 kg/cm^2. for 7 min.) at 300 or 330 min. after insemination. The ratio of normal embryos to total eggs used was only 7.2%.

3326 eggs of masu salmon were irradiated with 5 to 10 x 10^4 rads of gamma rays and were inseminated with normal sperms. Eggs were maintained at 10° C and subjected to hydrostatic pressure (650 kg/cm^2, for 6 min.) at various times (during 210 to 370 min. after insemination). The best results, 7.2% of treated eggs were normal at eyed stage, were obtained when eggs were treated at 310 min. after insemination. A total of 53 normal androgenetic diploids were produced and 46 of them hatched out. They were viable and grew up normally.

Flow Cytometry vs. Nuclear Area: Analysis of Induced Polyploidy in *Ctenopharyngodon idella* x *Aristhichtys nobilis* Hybrids

S. K. Allen, Jr. and J. M. Myers, University of Washington, Seattle

The identification of alterations in ploidy has been hindered by the lack of an exact, expedient, and economical assay. Putative triploid hybrids between *C. idella* and *A. nobilis* (J.M. Malone and Sons, Lonoke, Arkansas) were selected as subjects due to the distinct morphological characteristics differentiating diploid from triploid hybrids and hybrids from either parent. Non-lethal blood samples were drawn from each fish in order to directly compare the current method for evaluating ploidy in these hybrids–erythrocyte nuclear volume–and a direct measure of DNA via flow cytometry.

To measure nuclear volume, blood was smeared, stained and photographed at 1000x; photomicrograph negatives were mounted in 35mm slide frames and projected onto a screen. Nuclear areas were traced and later retraced on a digitizer. Blood cells from flow cytometry were stained in the DNA fluorochrome DAPI and the fluorescence from 1000–5000 cells quantified for each fish. Measurements from putative triploid hybrids were compared to those obtained from diploid *C. idella*.

The flow cytometry assay provided an unequivocal determination of ploidy. We were unable to identify distinct diploid/triploid groups using the nuclear volume technique. The range of triploid individual means differed greatly depending on the method used; flow cytometry provided an average twelvefold reduction in the range. The correlation between relative DNA contents (RDC's) obtained by both methods was .776. When morphologically derived RDC's were compared to those of either method, correlations of .997 (flow cytometry) and .762 (nuclear volume) were obtained. A comparison of cost showed flow cytometry, \$3.54/sample, to be 20% less costly than the nuclear volume technique, \$4.28/sample. The cytometry analysis for the 55 samples examined was completed in 5 hours, while erythrocyte measurements took 2 working days.Cytofluorometric analysis is recommended as the most efficient and definitive method for ploidy analysis.

Comparison of the Reliability of a Coulter Counter with a Flow Cytometer in Determining Ploidy Levels in Pacific Salmon

O. W. Johnson and P. Rabinovich, University of Washington; and F. M. Utter, National Marine Fisheries Service, Seattle

The reliability for differentiating ploidy levels of an ICP-22 Flow Cytometer using a DNA fluorescent dye and a Coulter Counter and Coulter Channelyzer measuring cellular volume was compared.

Fish from each of four groups of Pacific Salmon—*Oncorhynchus tshawytscha; O. gorbuscha;* and two groups of hybrids *O. tshawytscha* (female)/*O. gorbuscha* (male) and *O. gorbuscha* (female)/*O. tshawytscha* (male)—which had been heat treated at fertilization to produce triploid fish were analyzed at 15 months of age to determine ploidy using both techniques. Results are discussed with the objective of using the Coulter Counter as an inexpensive and more commonly available alternative to the flow cytometer.

Studies on Salt Water Adaptability and Sexual Maturity of Triploid Pacific Salmon

O. W. Johnson, F. M. Utter, W. W. Dickhoff, and R. N. Iwamoto, University of Washington, Seattle*

**National Marine Fisheries Service*

Fish from four groups of Pacific salmon which had been heat treated at fertilization to produce triploid fish, and whose ploidy levels had been monitored throughout development, were introduced into salt water. Blood samples were taken at two week intervals prior to and following the introduction of the fish into salt water, to analyze estradiol and thyroid hormone levels and to monitor ploidy. Relative sexual development of diploid and triploid individuals was also examined at regular intervals.

Significant differences in hormone levels, sexual development, and mortality were observed between diploid and triploid fish, suggesting that alternation in ploidy levels has a profound effect upon development and salt water adaptability.

Methods and Implications of Inducing Androgenesis in Salmonids

Jim Parsons and Gary Thorgaard, Washington State University, Pullman

Heterotic effects obtained in progeny generated from crosses between inbred lines have proved to be extremely useful in agriculturally important crop species. Work with crosses between inbred lines of rainbow trout have shown similar success, but are limited due to the long process inherent in the conventional methods for production of inbred lines.

Androgenesis, production of individuals with both chromosome sets from the male parent, could provide an efficient method for the rapid generation of inbred lines. In addition, androgenesis could also be used to restore diploid individuals directly from cryopreserved sperm (sperm banks), and may help in gaining a better understanding of salmonid genetics. Androgenesis involves inactivation of the egg nucleus with radiation, fertilization with normal sperm, and subsequent blockage of the first cell division of the developing haploid embryo to produce a diploid having both chromosome sets from the male parent.

Techniques for successful inactivation of the egg nucleus have been characterized. Current work on blockage of the first cell division shows promise, but appears to be associated with high mortality rates.

Enhancing the Reproductive Potential of Captive Coho Salmon Broodstock

R. N. Iwamoto and W. K. Hershberger, University of Washington; and Carlin McAuley, Domsea Farms, Inc., Bremerton, WA

An integrated program has been designed and implemented to provide coho salmon broodstock for the controlled environment and production requirements of marine net-pen culture. While substantial progress relative to non-selected controls has been achieved in sub-adult performance traits (fresh- and seawater growth, smoltification, and seawater survival), the management of adults under captive conditions and their requirements for subsequent survival and maturation require definition. The parameters currently used to evaluate the status of broodstock include physiological measurements, effectiveness of prophylactic disease treatments, and determination of appropriate holding conditions. Results of the use of these techniques and genetic manipulation are presented.

Development of Biannual Spawning Behavior in a Rainbow Trout Population

Harold L. Kincaid, U.S. Fish and Wildlife Service, Kearneysville, WV

During December 1974, hybrid matings between males from normally fall-spawning (October) strains and females from normally winter-spawning (January) strains were produced. The F_1 progeny matured as two year fish in December, 1976 as expected. In June 1976, five females were found mature after a 6-month spawning interval. A selection and hybridization program was established that has resulted in an increased frequency of fish exhibiting the biannual spawning trait since 1976. Data from the 1980 year class through three years of age show that 96.9, 86.5, and 93.0% of the fish spawned at 2.0, 2.5, and 3.0 years of age. Fifty-eight individually identified fish were evaluated after the 2.5 year spawning period yielding 36 females (75% mature at both periods), 11 males (64% mature at both periods), and 5 fish not mature at either spawning period.

Egg quality measured in terms of egg hatchability during the May-June period has been low, ranging from 0 to 56% in 56 lots spawned in 1982. Fecundity of the fish at 2.0, 2.5, and 3.0 years was 2197, 2153, and 3623, respectively. Management implications of biannual spawning is discussed.

The Consequences of Maintaining Families in a Commercial Breeding Program for Atlantic Salmon

G. F. Newkirk, Dalhousie University, Halifax, Nova Scotia, and S. Merrill-Stavostrand, IMA Aquatic Farming Ltd, Argyle Head, Nova Scotia, Canada

In a study designed to evaluate a breeding program for Atlantic salmon, *Salmo salar*, a stock establishment and evaluation program was started in a commercial hatchery. Stock was maintained as separate full sib families. The facilities were designed to hold the eggs and fry in separate tanks until the fingerlings were large enough to be given a family brand and mark. The resulting loss in growth due to this procedure has meant a serious loss in yield of 1 + smolts compared to contemporaneous fish handled according to conventional hatchery procedures. The advantages of family structure in a breeding program is discussed in terms of control of pedigree and obtaining desirable genetic parameters. In the particular case studied it seems the disadvantages in terms of loss of population are too great and an alternative approach is recommended.

Evaluating the Consequences of Reproduction in Complex Salmonid Life Cycles

Hal Caswell, R. J. Naiman, Roderick Morin, Woods Hole Oceanographic Institution, Woods Hole, Massachusetts

The demographic methods used in modern life history theory were originally developed for human populations. The life cycles of many organisms, however, differ significantly from that of man, and thus are difficult to analyze by traditional means. In salmonids, these differences include the partition of the life cycle between two distinctly different environments, plasticity in growth which makes maturation an ill-defined function of age, the existence of multiple reproductive pathways, and the divergence of the life cycles of males and females. Recent developments in evolutionary demography now make it possible to analyze such complex life cycles in great detail. This paper outlines these developments in general terms, and presents preliminary results of their application to the Atlantic salmon *(Salmo salar)* L. of the Matamek River, Quebec.

In particular, we focus on the problem of evaluating the relative contribution to population dynamics of reproduction of different stages in the life cycle. We show that quantitative estimates of these contributions can be made, and that they depend on both the stable stage distribution and the reproductive value distribution of the population. In an evolutionary perspective, this provides a measure of the selective importance of the different reproductive pathways. Recent changes in the reproductive life history of the Atlantic salmon (increased incidence of precocious maturation by males and of spawning by one sea year females) may be explicable on this basis.

Sex Control in Rainbow Trout for Mariculture in British Columbia

I. I. Solar, Ministry of Environment; E. M. Donaldson and G. A. Hunter, Department of Fisheries and Oceans, West Vancouver, British Columbia, Canada

Methods of hormonal sex control and chromosomal manipulation have been considered in the last decade as means of increasing both the quality and the quantity of cultured salmonids. In British Columbia we are studying the application of these techniques to the production of all female and sterile stocks of rainbow trout for aquaculture.

The effect of oral administration of methyltestosterone on sex reversal and sterilization of rainbow trout and the effect of heat shocks on the induction of triploid fish have been investigated. The administration of 17α-methyltestosterone at concentrations between 1 and 100 mg/kg of food for periods up to 120 days following swim-up produced male-female ratios substantially different from the control. No dose-related mortality during the period of treatment was observed.

Heat shocks were applied to rainbow trout eggs for 10 min. starting at different times after fertilization and at temperatures ranging from 24° to 30°C. Analysis of blood samples by flow cytometry revealed triploid induction ranging from 18 to 100% depending on the temperature and timing of the treatment. Survival from fertilization to 60 days and growth of the treated groups were lower than the control.

Environmental Aspects

Session Leader
Rolland Billard

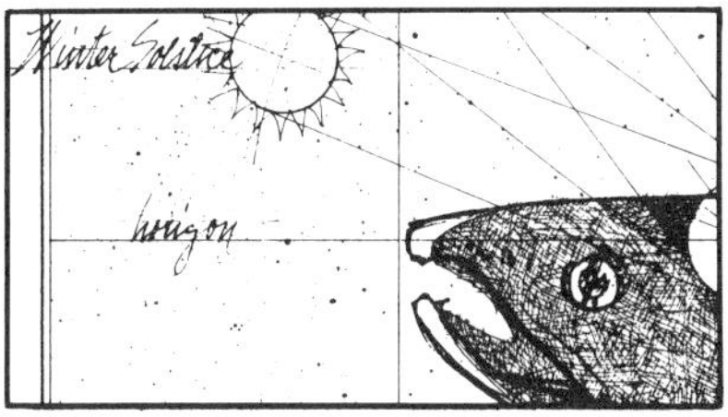

ENVIRONMENTAL FACTORS IN SALMONID CULTURE AND THE CONTROL OF REPRODUCTION

Rolland Billard[1]

Summary: A number of environmental factors affect reproductive function in fish. The effects of these factors vary with the species and developmental stage. The most visible factors implicated are photoperiod and temperature. Photoperiod acts on the establishment of puberty, gametogenesis and ovulation. Temperature, being an environmental signal, has similar effects but they are usually short-termed; this signal coordinates various stages (gamete release, fertilization, embryogenesis) with the most favorable conditions for the progression of each stage. Temperature also acts on puberty and gametogenesis. Other factors that are also important are a) salinity which affects puberty, gametogenesis, the quality and survival of gametes, fertilization and embryogenesis and b) diet which affects puberty and gametogenesis. The social environment may also act on puberty, gametogenesis and the efficiency of gamete release.

Anthropic factors (change in water flow rate, pollution, river navigation) also affect reproduction, usually in a negative way. Environmental factors often have an indirect effect via sensorial functions and the central nervous system but may also directly affect gonads, gametes and eggs. Some rhythmic-type factors in the environment may also entrain internal cycles.

Introduction

Fish reproduction in temperate and even in tropical zones (Lowe-McConnell, 1979) is seasonal and depends on diverse environmental factors. The most evident factors controlling reproduction differ according to the geographical zone, the species and the stage of reproduction. These factors have been considered to have a role in the determination of the timing of the breeding season. They act at different levels, as, for example, temperature and food on the gonad, and photoperiodic change and social interaction on the central nervous system.

It is not clear whether salmonids respond to night and day by means of a photoperiodic clock, based on a circadian system and exhibiting an annual reproductive cycle, or whether the cycle results from a true circannual rhythm entrained by some environmental factors. Daily or annual light and temperature changes are the most obvious entrainers but other changes may be identified in the environment which act as putative "zeitgeber." For practical application, it is necessary to know the relationship between environmental factors and reproduction because a) we can predict reproductive success from the record of pre-

[1] I.N.R.A., Laboratoire de Physiologie des Poissons, France

vailing environmental conditions and valuable wild species of fish can be managed better (Wootton, 1982) and b) the main environmental factors can be manipulated in fish farming to control some critical phases of the reproductive cycle (Billard and Breton, 1982; Billard, 1980, 1982; Bromage et al., 1982).

The role of the environment on reproduction has been reviewed recently (Wootton, 1979, 1982; Scott, 1979; Baggerman, 1980; Gerking, 1980, 1982; Billard et al., 1981a; Colombo et al., 1982; Bromage et al., 1982; Hokanson, 1984). Therefore, the scope of this paper has been limited to a review of the effect of some environmental factors on puberty, gametogenesis, spawning, gamete survival and embryogenesis. The possibility of using these factors to control reproduction is discussed.

Puberty

Information on the environmental factors involved in the initiation of puberty in salmonids is limited. Growth rate, probably a major factor (Alm, 1959), is highly dependent on food availability and on photoperiod, temperature and social interaction as well as on other factors. A better diet or overfeeding advances the age of puberty in rainbow trout (Kato, 1975, 1978) and arctic char (Runnstrom, 1951). Higher temperature also advances the age of puberty (Titarev, 1975), probably via a faster growth rate. Contradictory information has been reported on the effects of photoperiod on the first reproductive cycle. Several papers report that various photoperiodic manipulations, including continuous light or darkness, have no effect on the first reproductive cycle (Henderson, 1963; Pyle, 1969; Bieniarz, 1973). However, Shiraishi and Fukuda (1966), Lundqvist (1980), Ericksson and Lundqvist (1980), and Bourlier and Billard (1984a) were able to retard the first reproductive cycle by using long day or constant light (see below). Contraction of the annual photoperiodic regime into a 6-month period advances puberty (ovulation) in rainbow trout (Whitehead et al., 1978). In the same species a decreasing photoperiod starting in April stimulates spermatogenesis, but only in fish which are already at a prepuberal stage (Magri, 1983). Most of these experiments indicate that photoperiod has only a limited effect on the onset of puberty.

The problem of puberty is more complex in migrating salmonids which exhibit smoltification instead of puberty and in which initiation, and sometimes completion, of the first cycle occurs in the sea in an environment different from fresh water (for review of the case of Atlantic salmon see Gardner, 1976). The mechanisms controlling puberty are not known. This considerably limits the possibility of preventing the phenomenon of precocious maturation which is more and more frequent in salmonid farming and is probably due to an overall improvement of rearing conditions and thus to accelerated growth.

Gametogenesis

Gametogenesis in male and female salmonids, excluding Pacific salmon, has one basic difference. In males, spermatogenesis is accomplished within one year, while in females the oogenetic cycle occurs over a period of 2 years or more, the gonad growing slowly for one (or several?) year(s) and rapid growth (vitellogenesis) occurring for only a few months in the last year of the cycle. The situation (often ignored because most work deals with the rapid-growth phase) is important because at least two successive cycles overlap, and manipulation in the last part of one cycle interferes with the first part of the next one.

In many salmonid species there is great individual variation in the onset of spermatogenesis and vitellogenesis; this is also true of the time of spawning. Spermatogenesis starts in the summer and vitellogenesis in the spring, although rapid growth occurs mainly in the summer and may continue during the winter in the case of spring-spawning strains of rainbow trout (see below).

Influence of Photoperiod and Manipulation of the Sexual Cycle

Advancement of the sexual cycle: Because spermatogenesis and vitellogenesis occur in summer, decreasing photoperiods were first considered to be the main environmental factors. In fact, various experiments have shown that gametogenesis occurs under various photoperiodic conditions. To advance the onset of vitellogenesis and spermatogenesis, four main combinations are used (Figure 1). They are listed below with recent references. Previous studies have been summarized by Htun Han (1977).

a) Contraction of natural photoperiodic changes into periods of less than 12 months. This was the first method used (Hoover and Hubbard, 1937); it has been reexamined recently by Buss (1980) in brown trout and Billard and Breton (1977), Whitehead et al. (1978), and Pohl et al. (1982) in rainbow trout. These experiments were carried out under constant temperature regimes. Contraction into a 6- or 9-month period resulted in a normal reproductive cycle in males. Contraction into a 3-month period resulted in erratic spawning (Pohl et al, 1982). The efficiency of spermatogenesis (measured by the GSI and analysis of spermatogenesis) was not changed by the photoperiod but only by temperature (Billard and Breton, 1977). In females, plasma GTH and vitellogenetic profiles (phosphoproteins and calcium), but not that of estradiol, were advanced in contracted groups (Bromage et al., 1982; Whitehead et al., 1978). Fecundity was slightly lower (Billard and Breton, 1977) and egg diameter was less (Pohl et al., 1982) in the contracted than in the control cycle. The fertility of the sexual products was normal (Buss, 1980; Pohl et al., 1982).

In coho salmon, McQuarrie et al. (1978, 1979) tried to advance spawning by contracting the annual photoperiodic cycle into 8 months, but using un-

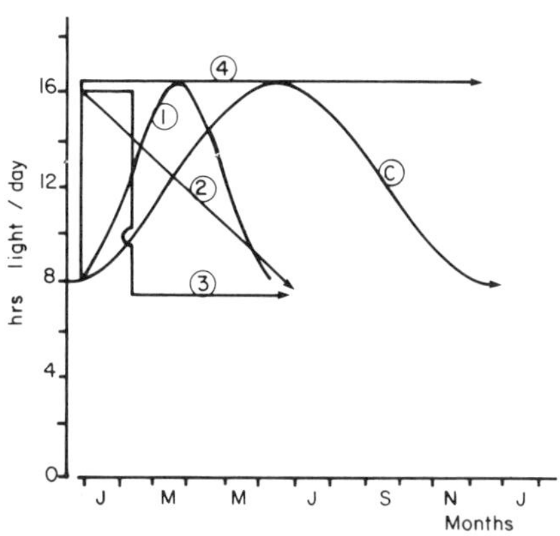

Figure 1. Various photoperiodic regimes used to advance ovulation in salmonids. 1: contracted regime; 2: advanced decreasing regime; 3: long days followed by short days; 4: constant photoperiod; 5: control; natural annual photoperiod.

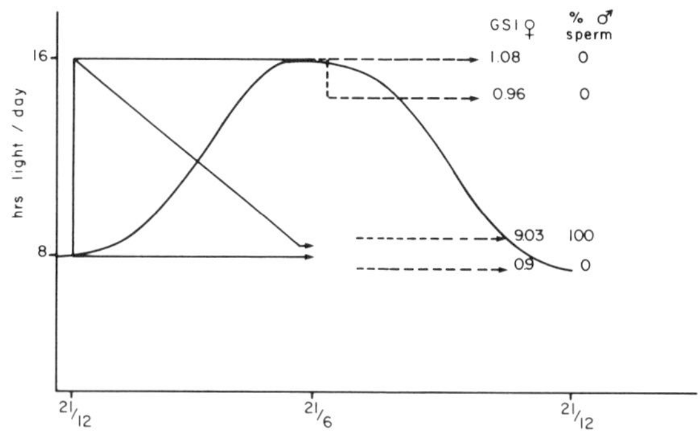

Figure 2. Effect of some photoperiodic regimes on the advancement of vitellogenesis (measured by the GSI) and spermiation (measured by the percentage of ripe males) in rainbow trout. Measurements were taken in June.

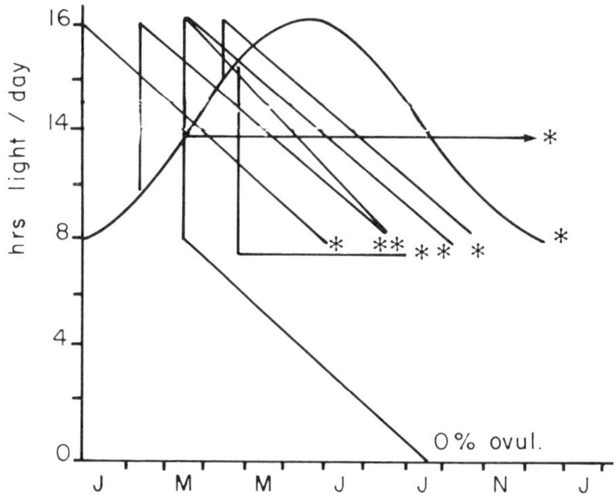

Figure 3. Effects of some photoperiodic regimes on the advancement of ovulation in rainbow trout: + indicates that ovulation occurred, and O, no ovulation

modified natural temperature. Spawning and probably vitellogenesis were advanced by 3 months but egg quality was slightly poorer than in the control group.

b) Advancement of the decreasing photoperiodic regime. The normal regime (~16L:8D → 8L:16D within 6 months) is usually advanced and starts between January and April. Experiments were carried out on rainbow trout (Billard and Breton, 1977; Breton and Billard, 1977; Breton et al., 1982; Billard et al, 1981a). In these conditions, with a 16L:8D regime starting at the winter solstice, vitellogenesis and spermatogenesis were initiated earlier than in the control and at the summer solstice 100% of the males were in spermiation; the GSI was increased in females (Figure 2). When the decreasing regime started later, after the normal rising regime had begun in February–April, ovulation as well as spermiation occurred around 6 months later. A similar decline of 8 hours from 8L:24D → 0L:24D did not result in any advancement (Figure 3). When the fish were kept under constant photoperiod 12L:12D or 8L:16D, there was also no advancement in comparison with the controls. All these experiments were carried out in the laboratory under constant temperature. This procedure was applied in rainbow trout and brown trout farms with naturally fluctuating temperatures (Breton et al., 1983) (Figure 4). In both cases, ovulation and spermiation were advanced while fecundity remained unchanged. Fertility was normal in brown trout which spawned in the fall at a temperature of less than 10°C. Ovulation in

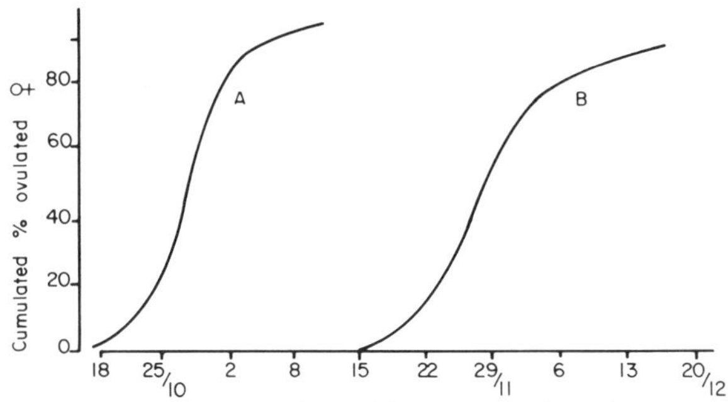

Figure 4. A. Advancement of ovulation in brown trout submitted to an advanced decreasing photoperiod starting on April 4 (16L:8D) and reaching 8L:16D on October 1. B. Ovulation curve in control females kept under natural photoperiod.

rainbow trout occurred in mid-summer when the temperature was too high for good post-ovulatory survival. For this technique to be valid, the females must be checked very frequently for ovulation and spring, well or cooled water must be used for incubation.

c) Long days followed by short days. This method is related to the previous one; the brood stock are first placed under long days (16L:8D, for instance) for 1 to 3 months, after which the regime is abruptly changed to short days (8L:16D). Ovulation occurs 2 to 3 months afterwards, depending on the temperature. This approach was used by Henderson (1963) and Bromage et al. (1982), who showed an advance in the profiles of GTH and estradiol secretion and by Fushiki (1979) and Takashima and Yamada (1984). When the normal rising regime was brought abruptly to short days in April, ovulation was also advanced (see Figure 3).

d) Constant photoperiod. In this case, the photoperiod remains constant for short days (8L:16D), long days (16L:8D) or continuous light (24L:0D). Contradictory results have been reported. (Pyle, 1969; Poston and Livingston, 1971, and Buss, 1980). Spermiation was advanced in male brook trout but ceased early. On the contrary, Bourlier and Billard (1984a) observed a delay in spermiation. Whitehead et al. (1978) and Bromage et al. (1982) using 12L:12D reported that reproduction occurred at the same time as the controls (see Figure 3). Skarphédinsson et al. (1982) and Bromage et al. (1984) indicated that ovulation and plasma testosterone levels show a 6-month cycle under a constant photoperiod of 18L:6D; fish spawned under 16L:8D. Constant long days starting in February–April advance spawning. Long days given later in the year delay spawning. The time of introduction of fish onto the modified photoperiod regime is the critical factor.

Delay of the sexual cycle: In sub-adult rainbow trout (first reproductive cycle), the onset of vitellogenesis and spermatogenesis is delayed when the fish are kept under long days or brought to constant light after the summer solstice; ovulation and onset of spermiation are also delayed by one or two months (Bourlier and Billard, 1984a,b). Similar results were reported previously in female (but not male) brook trout kept under constant light or dark (Poston and Livingston, 1971; Pyle, 1969). Long days or constant light given after July delays maturity in the males and females of four salmonid species (Shiraishi and Fukuda, 1966) and in Atlantic salmon (Lundqvist, 1960), but not in rainbow trout (Bieniarz, 1973).

The problem of obtaining more than one spawning per year: There is now increasing evidence that some fish may have two reproductive cycles in one year. This has been reported to have occurred spontaneously in a fish farm in California (Hume, 1955) and in Japan (Nomura, personal communication) and has been demonstrated experimentally in the laboratory after light manipulation: fluctuating photoperiod (Breton et al., 1982) and 18L:6D (Skarphédinsson et al., 1982). At present, strains having two spontaneous reproductive cycles under natural environment are currently established in Japan (Aida et al., 1984) and in the USA (Kincaid, 1984).

Conclusions: Seasonal reproduction in salmonids does not seem to be a true circannual rhythm since at least two cycles may occur in a 12-month period. Two cycles have also been observed under constant photoperiod (18L:6D) and temperature (i.e. free running in the absence of zeitgeber). Gametogenesis in salmonids is more likely initiated and entrained by such signals as day length, and possibly changes in temperature, but no experiment has been designed to demonstrate this. This initial change seems to occur within a precise range of day length (or night length), i.e. 16L:8D → 8L:16D, since a change of similar amplitude (8L:16D → 0L:24D) has no effect. In addition, abrupt or progressive changes have the same effect. Ovulation occurs 3 to 4 months after an abrupt change; this period, plotted back in the case of progressive decline, shows that the initial effect occurs when more than 12L are used. How salmonids measure changes in night or day length is not known. Observations by Bromage et al. (1982) would favor the entraining system of circadian oscillations rather than the hour-glass system, although the hypothesis of a circadian photosensitive phase shifting between 16L and 10L during gametogenesis has not been confirmed by Billard et al. (1981b) and Bromage (personal communication).

Even though control of the sexual cycle by light manipulation can be practically applied, actual knowledge of the seasonality of gametogenesis is not fully understood. Contradictory reports in the literature may result from differences in the experimental conditions, e.g. cycle used (first or second), onset of darkness (light switched off abruptly or progressively), differences in light intensity and temperature regime. The role of temperature itself or its interaction with light may have been underestimated; they may both have an initiatory effect like

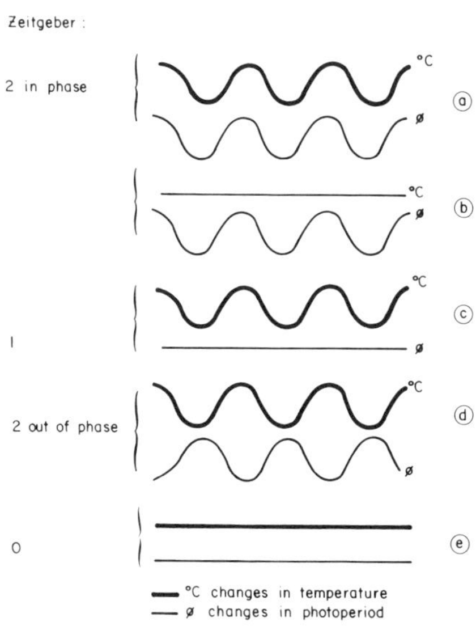

Figure 5. Various combinations of light and temperature annual changes in some experiments of photoperiodic manipulation (see text).

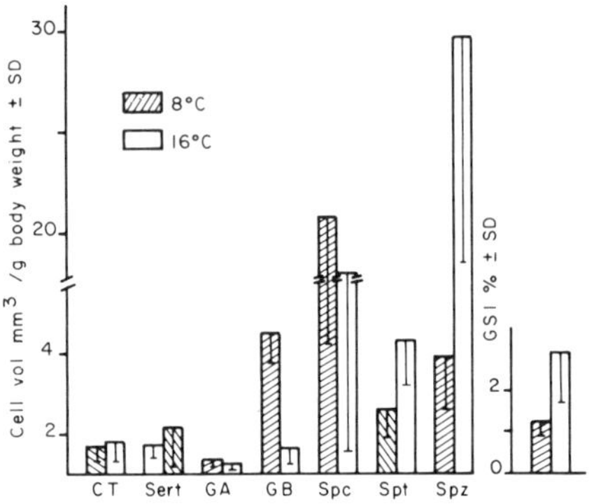

Figure 6. Effect of constant rearing temperatures (8° and 16°C) on the efficiency and speed of spermatogenesis in rainbow trout bred from March 4 to July 2 under decreasing photoperiod (16L:8D → 8L:16D). n = 10. Spermatogenesis is more advanced and the GSI is higher at 16 than at 8°C. (From Billard and Breton, 1977)

zeitgeber. The different situations encountered in the experiments reported above (Figure 5) were highly diversified:

a) temperature and light changes were in phase; this is the normal situation in nature;

b-c) temperature or photoperiod was constant and only one zietgeber was present. (b) is illustrated by the experiment of Bromage et al. (1982) and Breton et al. (1981), and (c) by Bieniarz (1973) who obtained normal cycles in total darkness or constant illumination. Temperature change may not be involved since Pyle (1969) obtained the same results under the same light conditions but at constant temperature, although the fish were in their first cycle;

d) temperature and photoperiodic changes were out of phase, affecting the quality of the eggs (see McQuarrie et al, 1978);

e) temperature and photoperiod were constant as in Bromage et al.(1982).

Influence of Temperature and Other Environmental Factors on the Sexual Cycle

In most of the works reported above, the photoperiod appears to be a major factor determining the process of gametogenesis, but temperature may intervene directly on the gonads controlling the speed and efficiency of this process (Figure 6). In northern zones, spawning is delayed until summer due to low temperature and ice cover. After spawning, the season for gonadal development for the next cycle may be too short, especially if there is little food available (see below). This would explain why gametogenesis may occur every two years in northern Canada (Moreau, personal communication). Temperature is an important factor in salmonids, especially at the end of the cycle. In spring-spawning strains, complete gonadal development does not occur in the fall unless the temperature exceeds 5°C (Goryczko, 1972). If the temperature is lower, gonadal development is retarded but the oocytes continue to grow in the winter at low temperature (Soivio, personal communication). Fall-spawning strains are found only in hatcheries with a warm-water supply (Islam et al., 1973; Hokanson, 1984).

Nutrition is another major environmental factor. The importance of food availability is often stressed by authors (reviews by Woodhead, 1960, 1978; Wootton, 1979, 1982; Billard, 1980), but experimental data are scarce in salmonids. Fecundity is usually higher when ample food is available to the females (Scott, 1962; Bagenal, 1969a; Billard and de Fremont, 1980). Egg size also tends to increase, except in Bagenal's study, but the oocytes were probably measured before vitellogenesis was completed. In two Russian works, changes in alevin weight have been related to the feeding regime of the parents (Pchelovodova, 1982; Efimova and Sentishcheva, 1982). The fertility of rainbow trout gametes and their survival at resorption were not changed by a reduced diet (Billard and de Fremont, 1980), but Bagenal (1969b) has shown that brown trout embryos hatched from larger eggs are more resistant to starvation. Under food restriction, females may remain in a stage of sexual stasis in the summer and miss one annual cycle (Kennedy, 1953; Wydosky and Cooper, 1966; Dutil, 1982; Moreau, personal communication). Temperature was shown to interfere with food availability in the medaka (Hirshfield, 1980).

The problem of egg quality in relation to food composition has not been thoroughly investigated in salmonids. Compounds such as carotenoids or essential fatty acids have not been clearly shown to contribute to better egg quality. In *Coregonus albula* the chemical composition of the eggs was related to the lake in which the females originated. In lakes with a very limited food supply, the females had a low growth rate and generated poor quality eggs (Kamler and Zuromska, 1979). In another study (Zawisza and Bakiel, 1970), other factors in the lake, such as depth, O_2 and temperature, had no effect on fecundity in *Coregonus albula*.

The effects of stress on gametogenesis are not well documented in salmonids. Negative social interaction due, for instance, to increased density acting via "crowding-factor" effects (see Colombo et al., 1982 for review) are not likely to occur in salmonids considering that the water is renewed very often or aerated. Pollution has been shown to adversely affect gametogenesis (see review by Billard et al., 1981a), fecundity in brook trout is decreased by low pH (Menendez, 1976) and by iron hydroxide (Sykora et al., 1975) and in rainbow trout by DDT (Macek, 1968). HCN at a low dose (0.01 mg/l of water) alters the process of secondary yolk deposition in the oocytes of rainbow trout (Lesniak and Ruby, 1982).

Spawning

Spawning includes the final steps of the cycle: oocyte maturation, ovulation and oviposition in females and spermiation and sperm release in males. Most experimental work on spawning in salmonids has only concerned oocyte maturation, ovulation and spermiation. Photoperiod does not seem to be an essential factor or cue for ovulation and spermiation, since spawning occurs within a wide photoperiodic range. Temperature is a much more important factor. Optimal spawning temperature ranges from 5.6° to 13°C (Leitritz and Lewis, 1976; Piper et al., 1982). Outside this range, adult and egg survival and ovulation are adversely affected. Maturation and ovulation in chinook salmon and rainbow trout are blocked at temperatures lower than 4°C (Leitritz and Lewis, 1976) and in brown trout between 2° and 4°C (Hokanson et al., 1973). Males were ripe at 1-2°C but spermiation may have started at a higher temperature. At temperatures above 13°C, ovulation occurs in rainbow trout but post-ovulatory egg survival in the body cavity is much shorter (see above and Figure 7), although better than *in vitro* (Billard and Gillet, 1981). In brook trout ovulation and spawning occur at 16°C and lower (Hokanson et al., 1973). The sperm fertilizing capacity of male rainbow trout is not modified when the trout are kept at 10°, 15° or 18°C (Figure

8). Hokanson et al. (1973) studying brook trout observed that 19°C was the maximal temperature at which the males were mature with motile spermatozoa. The spermiation yield decreased at higher temperatures (Figure 9), and a sudden rise in temperature (10° to 18°C in 2 days) resulted in a significantly lower volume of milt collected when compared with controls kept at 10°C, although gonadotropin secretion was stimulated (Figure 10).

11-Ketotestosterone (11KT), which appears to be the major androgen involved in spermiation (Fostier et al., 1982, 1983), may have been glucuronidated by the highest temperature; Kime (1979) has shown *in vitro* that glucuronidation of 11KT increases with the temperature. It is supposed that temperatures in the male and female act at the level of the gonad, but an action at the level of the central nervous system cannot be excluded since the process of ovulation is accelerated by the injection of pituitary extract in brown trout kept at a temperature lower than 3-4°C (Billard, unpublished data).

The social environment is also important. The presence of sexual pheromones originating from the ovary and exciting conspecific males has been demonstrated in salmonids as in other species (for review see Liley, 1982). The exis-

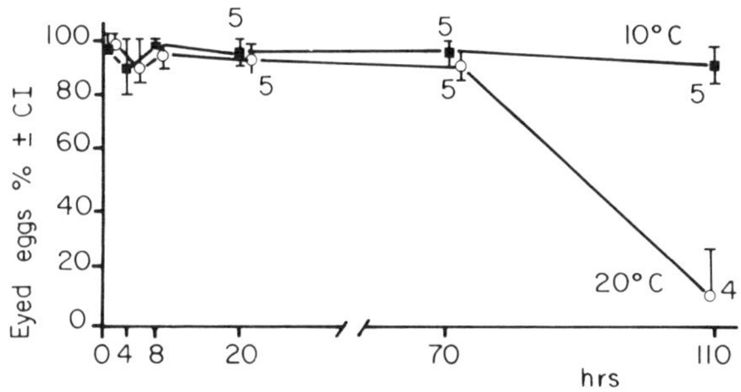

Figure 7. Changes in ovum fertilizability in the body cavity of female rainbow trout kept at 10° and 20°C for 110 hours after ovulation. (From Billard and Breton, 1977)

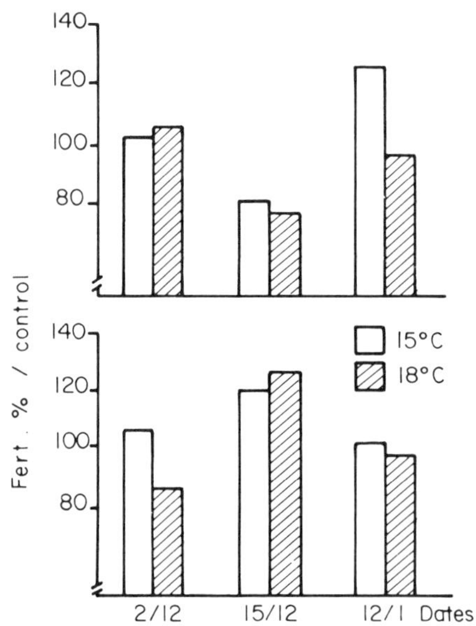

Figure 8. Fertilizing ability of sperm collected from males kept at 15° and 18°C from October 29 and expressed in percentage of the values of the control group kept at 10°C. (From Billard and Breton, 1977; also see Figure 9)

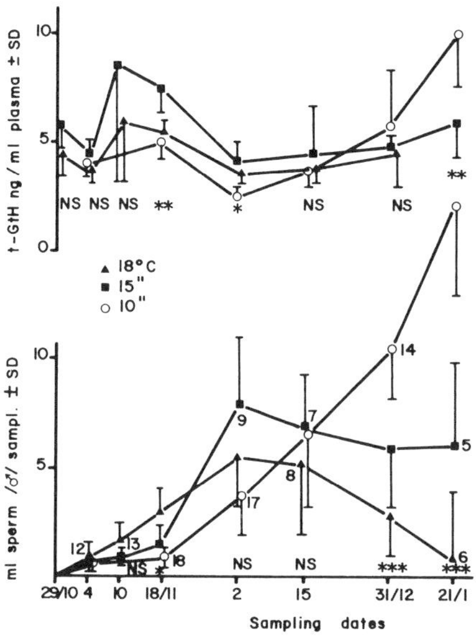

Figure 9. Changes in tGTH level and volume of sperm collected (ml/male/sampling) in rainbow trout males kept at 10°, 15° and 18°C from October 29. Fertilizing ability is shown in Figure 8.

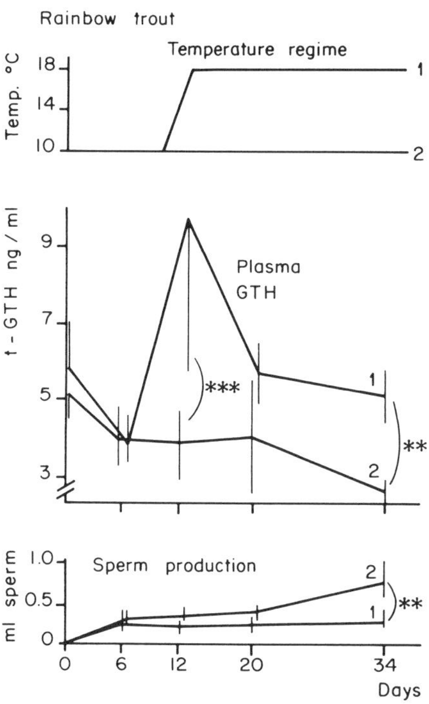

Figure 10. Changes in tGTH level and volume of sperm collected (ml/male/sampling) in two groups of males (n 10) exposed to two different thermal regimes (top graph). One group was kept at 10°C and in the other the rearing temperature was raised to 18°C. (From Billard and Breton, 1977). Day 0 = October 4. Values ± SD

tence of chemical signals originating from males and acting on females is less clear. Elucidation of the role of sexual pheromones on the final maturation state in conspecific individuals would contribute to good brood stock management at spawning time.

As far as reproduction and spawning are concerned, domesticated brook trout seem less sensitive to agression than wild ones (Vincent, 1980). Cortisol, which rises after drastic environmental change or aggression, has also been shown to increase at spawning time (Pickering and Christie, 1981). The meaning of such a rise is not clear. Cortisol has been shown to increase oocyte sensitivity to GTH and 17α-hydroxy-20β-dihydroprogesterone in rainbow trout (Jalabert, 1976), but better knowledge of the timing of such a rise is necessary before we can conclude that cortisol plays a functional role in ovulation and that stress does not harm this process.

Among other environmental factors influencing spawning are changes in water flow rate which probably intervene as timers of migration and spawning (Dodge and McCrimmon, 1971; Hokanson, 1984). This flow rate factor is often ignored in fish farms, but providing the brood stock with an adequate flow rate may help to keep it in good condition and to synchronize or stimulate ovulation or spermiation.

Gametes, Fertilization and Embryogenesis

The influence of some environmental factors such as temperature and diluent composition on gametes and ways of improving their management have already been reviewed in this volume (Billard, 1984). Temperature also has a considerable influence on embryogenesis. In fact, thermal requirement is much more strict during embryogenesis than during any other developmental stage in salmonids. The temperature range for incubation is 2.7-15.6°C in chinook salmon and 4-13°C in rainbow trout. Even within this range, temperature perturbation may result in changes in vertebral number (Lindsey and Arnason, 1981). Thermal shock is deleterious to the embryo but its effects vary with the stage of development. Sharp drops in temperature injure embryo cleavage, and spawning is completed before the water temperature falls below 4°C (chinook salmon). Dodge and MacCrimmon (1971) showed that when spawning in rainbow trout occurs at 0.3-2°C, embryonic mortality is very high (see review by Hokanson, 1984). In Atlantic salmon, embryonic mortality is higher and the fry are smaller when incubation is carried out at 12°C rather than at 8° or 10°C (Gunnes, 1979). Early stages are more sensitive to high temperatures than eyed-stages in brook trout (Hokanson et al., 1973).

The oxygen requirements of salmonids depend on developmental stage, illumination and even the rate of O_2 saturation of the medium (Hamor and Garside, 1975, 1979). These same authors estimated total O_2 consumption in Atlantic salmon during incubation as 28 g in O_2-saturated water. Oxygen deficiency during incubation results in poorer post-hatching growth and at emergence the alevins are smaller than when under optimal conditions of oxygen supply (brown trout and rainbow trout: Winnicki, 1967; lake char: Balon, 1980).

Fish embryos are very sensitive to changes in the physico-chemical quality of water; low pH, which is now a major environmental concern, is very toxic for embryos. After continuous exposure to acidic water, the lower sub-lethal limit of Atlantic salmon embryo is pH 5.0 and hatching is more difficult at pH 4.0-5.0. Embryonic sensitivity to low pH depends on the stage of development; at the early cleavage stage, the lower lethal limit is about pH 3.6 and just before hatching it is 3.0-3.1 (Daye and Garside, 1975, 1977, 1980a). Embryos are less sensitive to an alkaline pH level (Daye and Garside, 1980b). Although protected by the zona radiata (Beattie and Pascoe, 1978), the ova, eggs and embryos are sensitive to a wide range of toxic products (Olson and Marking, 1975; McKim, 1977; Wales, 1979; Gillet and Roubaud, 1983). The toxic effects of micropollutants are increased by temperature (Billard and Billet, 1981). Other factors interfering with gametes and incubation which the fish farmer should be aware of are:

Light intensity: Salmonid eggs require relative obscurity; a light intensity exceeding 300 lux has negative effects on rainbow trout eggs (Bieniarz, 1973). Atlantic salmon eggs are sensitive to white, yellow and UV light and astaxanthin has no protective effect according to Torrissen and Torrissen (1981). Light also interferes with hatching. Briefly exposing (1-2 hours) prehatching embryos of Atlantic salmon to a light intensity of 300 and 450 lux advances hatching by 3-4 days and the embryos are bigger compared with the controls kept in darkness (Melnikova, 1982). This seems to be a useful way to synchronize, and possibly predict, hatching.

Mechanical shock: Salmonid embryos are sensitive to mechanical shock at various developmental stages. Immediately after fertilization (cortical reaction) and during gastrulation 5-10 days later, embryos are sensitive to agitation (Smirnov, 1955; Billard, 1977; Carpentier, 1977; Jensen, 1981; Ievleva, 1967; Jensen and Alderdice, 1983). However, experimental results vary widely probably due to individual variation (the eggs of some females show exceptional resistance or are very sensitive to agitation) and to the nature of the agitation; it seems that when the position of the egg is changed (during siphoning, for example), mortality is higher (Hata, 1927, 1929; Marcel, 1981).

Maternal factors: Embryogenesis is also affected by maternal factors. Small eggs give embryos with fewer myomeres (Garside and Fry, 1959) and organochlorinated pesticide given to females during vitellogenesis gives a reduced percentage of developing embryos at the eyed-stage (Billard, 1978).

Conclusion

Temperature appears to be one of the most important environmental factors governing spawning activity, gamete viability and embryonic survival. The main practical applications of this fact have permitted us to prevent mortality during fertilization and incubation and to delay hatching time by carrying out incubation at lower temperatures (see Maddock, 1974). The latter method requires a water-cooling system which is expensive; a less expensive alternative is to carry out incubation in air with high humidity.

In hatcheries it is important to offer the embryo optimal environmental conditions and water quality. Inadequate temperature, oxygen, pH, salinity or the presence of toxic products in the water and deleterious therapeutic treatment may not kill the eggs but produce poor-quality fry with poor growth performance.

The brood stock as well should be put in optimal environmental conditions, not only to obtain a maximal number of eggs but also to ensure their quality.

Finally, the control of reproduction leading to the qualitative and quantitative production of alevins at a desired time may be achieved in part by combining the numerous environmental factors involved in each step of the cycle, that is, at gametogenesis, spawning and embryogenesis. This approach may also be associated with other fields such as genetics and endocrinology.

Acknowledgments

We wish to thank Alice Daifuku for the English presentation of the manuscript and Dr. M.H. Magri and N. Bromage for helpful comments on the paper.

References

Aida, K., Lou, S.W., Hanyu, I., 1984. Reproductive activity of a twice-annually spawning strain of rainbow trout. Int. Symp. Salmonid Reproduction, Seattle, Washington.

Alm, G., 1959. Connection between maturity, size and age in fishes. Rep. Inst. Freshwat. Res. Drottningholm 33:1-145.

Bagenal, T.B., 1969a. The relationship between food supply and fecundity in brown trout *Salmo trutta* L. J. Fish Biol. 1:167-182.

Bagenal, T.B., 1969b. Relationship between egg size and fry survival in brown trout *Salmo trutta* L. J. Fish Biol. 1:349-353.

Baggerman, B., 1980. Photoperiodic and endogenous control of the annual reproductive cycle in teleost fishes. In: M.A. Ali (editor), Environmental Physiology of Fishes, Plenum Pub. Corp., New York. pp. 533-567.

Balon, E.K., 1980. Early ontogeny of the lake charr, *Salvelinus* (Cristivomer) *namaycush*. In: E.K. Balon (editor), Charrs, Junk Publ., The Hague, Netherlands. pp. 485-562.

Beattie, J.H. and Pascoe, D., 1978. Cadmium uptake by rainbow trout, *Salmo gairdneri* eggs and alevins. J. Fish Biol. 13:631-637.

Bieniarz, K., 1973. Effect of light and darkness on incubation of eggs, length, weight and sexual maturity of sea trout (*Salmo trutta* L.), brown trout (*Salmo trutta* fario L.) and rainbow trout (*Salmo irideus* Gibbons). Aquaculture 2:299-315.

Billard, R., 1976. Effet de l'agitation des ovules et des oeuf de truite arc-en-ciel sur la fécondabilite et le début du développement embryonnaire. Bull. Fr. Pisci. 262:5-11.

Billard, R., 1978. Effect of heat pollution and organo-chlorinated pesticides on fish reproduction. In: C.E.C., Final reports on research sponsored under the first environmental research programme, Brussels. pp. 265-276.

Billard, R., 1980. Les possibilitiés de contrôle de la reproduction chez les poissons par modulation des facteurs de l'environment et manipulations hormonales. Cahiers Lab. Montereau 10:11-32.

Billard, R., 1982. The control of fish reproduction in aquaculture. World Conf. on Aquaculture, Venise, September 1981.

Billard, R., 1984. Artificial insemination in salmonids. Int. Symp. Salmonid Reproduction, Seattle, Washington.

Billard, R. and Breton, B., 1977. Sensibilité à la température des différentes étapes de la reproduction chez la truite arc-en-ciel. Cahiers Lab. Montereau 5:5-24.

Billard, R. and Breton, B., 1982. Control of reproduction and fish farming. In: B. Loft (editor), IXth Conf. Eur. Soc. Comp. Endocrinol., Hong Kong.

Billard, R. and de Frémont, M., 1980. Taux d'alimentation pendant la gamétogenèse et performance de reproduction chez la truite fario. Bull. Fr. Pisci. 279:49-56.

Billard, R. and Gillet, C., 1981. Vieillissement des ovules et potentialisation par la température des effets des micropolluants du milieu aqueux sur les gamètes chez la truite. Cahiers Lab. Montereau 12:35-42.

Billard, R., Bry, C., and Gillet, C., 1981a. Stress, environment et reproduction in teleost fish. In: A.D. Pickering (editor), Stress and Fish, Academic Press, London, New York. pp. 185-208.

Billard, R., Reinaud, P., and Le Bren, P., 1981b. Effects of changes of a photoperiod on gametogenesis in the rainbow trout (*Salmo gairdneri*). Reprod. Nutr. Devel. 21:1009-1014.

Bourlier, A. and Billard, R., 1984a. Delay of gametogenesis and spawning by constant illumination of rainbow trout (*Salmo gairdneri*) during the first reproductive cycle. Submitted for publication.

Bourlier, A. and Billard, R., 1984b. Delayed gametogenesis and spawning in rainbow trout (*Salmo gairdneri*) kept under permanent light during the first and second reproductive cycle. Aquaculture, submitted for publication.

Breton, B. and Billard, R., 1977. Effects of photoperiod and temperature on plasma gonadotropin and spermatogenesis in the rainbow trout *Salmo gairdneri*. Ann. Biol. anim. Bioch. Biophys. 17:331-340.

Breton, B., Zohar, Y., and Billard, R., 1982. Photoperiod and gonadotropin control of the reproductive cycle in the female rainbow trout. In: B. Loft (Editor), IXth Int. Soc. Comp. Endocrinol., Hong Kong.

Breton, B., Maisse, G., and Lemenn, E., 1983. Contrôle photopériodique de la saison de reproduction en salmoniculture : une expérience piolte en Bretagne. Bull. Fr. Pisci. 288:35-45.

Bromage, N., Whitehead, C., Elliot, J., Breton, B., and Matty, A., 1982. Investigations into the importance of daylength on the photoperiodic control of reproduction in the rainbow trout. In: C.J.J. Richter and J.H.Th. Goos (Comp.), Reproductive Physiology of Fish, PUDOC Wageningen. pp. 233-236.

Buss, K., 1980. Photoperiod control for brood trout. Manipulating the spawning season to meet production requirements. Aquaculture Magazine, Jan-Feb., 45-48.

Carpentier, P., 1977. Essais de conservation, de transport des gamètes et des oeufs fécondés et étude des conditions optimales de résorption vésiculaire chez les salmonidés. Dipl. Etudes Approf., Univ. Paris VI, 21 pp.

Colombo, L., Belvedere, P.C., Marconato, A., and Bentivegna, F., 1982. Pheromones in teleost fish. In: C.J.J. Richter and J.H.Th. Goos (Comp.), Reproductive Physiology of Fish, PUDOC Wageningen. pp. 84-94.

Daye, P.G. and Garside, E.T., 1975. Lethal levels of pH for brook trout, *Salvelinus fontinalis* (Mitchill). Can. J. Zool. 53:639-641.

Daye, P.G. and Garside, E.T., 1977. Lower lethal levels of pH for embryos and alevins of Atlantic salmon, *Salmo salar* L. Can. J. Zool. 55:1504-1508.

Daye, P.G. and Garside, E.T., 1980a. Structural alterations in embryos and alevins of the Atlantic salmon, *Salmo salar* L. induced by continuous or short-term exposure to acidic level of pH. Can. J. Zool. 58:27-43.

Daye, P.G. and Garside, E.T., 1980b. Development, survival and structural alterations of embryos and alevins of Atlantic salmon, *Salmo salar* L. continuously exposed to alkaline levels of pH, from fertilization. Can. J. Zool., 58:369-377.

Dodge, D.P. and MacCrimmon, H.R., 1971. Environmental influences on extended spawning of rainbow trout (*Salmo gairdneri*). Trans. Am. Fish Soc. 2:312-318.

Dutil, J.D., 1982. Periodic changes in the condition of the Atlantic charr (*Salvelinus alpinus*) of the Nauyuk lake system. Ph.D. Univ. Winnipeg, Manitoba, 149 p.

Efimova, T.A. and Sentishcheva, S.V., 1982. Influence of value of daily ration on quality of the rainbow trout spawners and their progeny. In: R.V. Kazokov, Les particularités biologiques des reproducteurs d'espèces de poissons à valeur économique élevée, Gos-Niorkh, Leningrad. pp. 84-89.

Ericksson, L.O. and Lundqvist, H., 1980. Photoperiod entrains ripening by its differential effect in salmon. Naturwissenschaften 67:202-203.

Fostier, A., Billard, R., Breton, B., Legendre, M., and Marlot, S., 1982. Plasma 11-oxotestosterone and gonadotropin during the beginning of spermiation in rainbow trout (*Salmo gairdneri* R.). Gen. Comp. Endocrinol. 46:428-434.

Fostier, A., Jalabert, B., Billard, R., Breton, B., and Zohar, Y., 1983. The gonadal steroids. In: W.S. Hoar, D.J. Randall and E.M. Donaldson (editors), Fish Physiology, vol. IX Reproduction, Academic Press, New York (in press).

Fushiki, S., 1979. Studies on the effect of long photoperiod in spring season on gonadal maturation of Ayu-fish *Plecoglossus altivelis* T. and S. Res. Rep. Shigo Prefectural Fisheries Experimental Station (Shigasuichi Kempo) 31:56.

Gardner, M.L.G., 1976. A review of factors which may influence the sea-age and maturation of Atlantic salmon *Salmo salar* L. J. Fish Biol. 9:289-327.

Garside, E.T. and Fry, F.E.J., 1959. A possible relationship between yolk size and differentiation in trout embryos. Can. J. Zool. 37:383-386.

Gerking, S.D., 1980. Fish reproduction and stress. In: M.A. Ali, Environmental Physiology of Fishes, Plenus, New York. pp. 569-587.

Gerking, S.D., 1982. The sensitivity of reproduction of fishes to stressful environmental conditions. In: C.J.J. Richter and J.H.Th. Goos (Comp.), Reproductive Physiology of Fish, PUDOC Wageningen. pp. 224-228.

Gillet, C. and Roubaud, P., 1983. Action du carbendazime, un fongicide antimitotique de synthèse, sur l'insémination de la truite arc-en-ciel et sur l'insémination et le développement précoce de la carpe commune. Water Res. (in press).

Goryczko, K., 1972. A change of the spawning season in rainbow trout (*Salmo gairdneri* Richardson). Roc nauk Roln Ser. H 94:57-68.

Gunnes, K., 1979. Survival and development of Atlantic salmon eggs and fry at three different temperatures. Aquaculture 16:211-218.

Hamor, T. and Gardise, E.T., 1975. Regulation of oxygen consumption by incident illumination in embryonned ova of Atlantic salmon *Salmo salar* L. Comp. Biochem. Physiol. 52:277-280.

Hata, K., 1927. On the influence of four kinds of vibration upon the eggs of *Oncorhynchus masou*. J. Imp. Fish. Inst. XXIII:74-78.

Hata, K., 1929. On the influence of the duration of time of vibration upon the development of fish-eggs. J. Imp. Fish. Inst. XXIV:112-123.

Henderson, N.E., 1963. Influence of light and temperature on the reproductive cycle of the eastern brook trout, *Salvelinus fontinalis* (Mitchill). J. Fish. Res. Board Can. 20:859-897.

Hirshfield, M.F., 1980. An experimental analysis of reproductive effort and cost in the Japanese medaka, *Oryzias latipes*. Ecology 61:282-292.

Hokanson, K.E.F., 1984. The effects of environmental factors on the measurement of temperature requirements for fish activities. E.P.A. Ed., in press.

Hokanson, K.E.F., McCormick, J.H., Jones, B.R., and Tucker, J.H., 1973. Thermal requirements for maturation, spawning, and embryo survival of the brook trout, *Salvelinus fontinalis*. J. Fish. Res. Board Can. 30:975-984.

Hoover, E.E. and Hubbard, H.F., 1937. Modification of sexual cycle of trout by control of light. Copeia 4:206-210.

Hume, L.C., 1955. Rainbow trout spawn twice a year. Calif. Fish and Game 41:117.

Htun-Han, M., 1977. The effects of photoperiod on reproduction in fishes. An annotated bibliography. Library Information Leaflet MAFF Lowestoft, 6, 30 pp.

Ievleva, M.Y., 1967. Resistance of developing eggs of sockeye salmon to mechanical action. Izv. Tikhookean. Nauchno-Issled. Inst. Rybn. Khoz. Okeanogr. 57:55-79. (Can. Transl. Fish. Aquat. Sci., no. 4617, 1980)

Islam, H., Nose, Y., and Yasuda, F., 1973. Eggs characteristics and spawning season of rainbow trout. Bull. Jap. Soc. Sci. Fish 39:741-751.

Jalabert, B., 1976. *In vitro* oocyte maturation and ovulation in rainbow trout (*Salmo gairdneri*), northern pike (*Esox lucius*), and goldfish (*Carassius auratus*). J. Fish. Res. Board Can. 33:974-988.

Jensen, J.O.T., 1981. Mechanical shock sensitivity of coho eggs. Proceedings of Workshops held in Seattle, Washington.

Jensen, J.O.T. and Alderdice, D.F., 1983. Changes in mechanical shock sensitivity of coho salmon (*Oncorhynchus kisutch*) eggs during incubation. Aquaculture 32:303-312.

Kamler, E. and Zuromska, H., 1979. Quality of eggs and production of *Coregonus albula* L. in three Masurian lakes. Po. Arch. Hydrobiol 26:595-623.

Kato, T., 1975. The relation between the growth and reproductive characters of rainbow trout, *Salmo gairdneri*. Bull. Fresh. Fish. Res. Lab. 25:83-115.

Kato, T., 1978. Relation of growth to maturity of age and egg characteristics in kokanee (*Oncorhynchus nerka*). Bull. Fresh. Fish. Res. Lab. 28:61-75.

Kennedy, W.A., 1953. Growth, maturity and mortality in the relatively unexploited lake trout, *Cristivomer namaycush*, of Great Slave Lake. J. Fish. Res. Board Can. 11:827-852.

Kime, D.E., 1979. The effect of temperature on the steroidogenic enzymes of the rainbow trout. Gen. Comp. Endocrinol. 39:290-296.

Kincaid, H.L., 1984. Development of biannual spawning behavior in a rainbow trout population. Int. Symp. Salmonid Reproduction, Seattle.

Leitritz, E., and Lewis, R.C., 1976. Trout and salmon culture. Fish Bull., 164.

Lesniak, J.A. and Ruby, S.M., 1982. Histological and quantitative effects of sublethal cyanide exposure on oocyte development in rainbow trout. Arch. Invironm. Contam. Toxicol. 11:343-352.

Liley, N.R., 1982. Chemical communication in fish. Can. J. Fish. Aquat. Sci. 39:22-35.

Lindsey, C.C. and Arnason, A.N., 1981. A model for responses of vertebral numbers in fish to environmental influences during development. Can. J. Fish. Aquat. Sci. 38:334-347.

Lowe-McConnell, R.H., 1979. Ecological aspects of seasonality in fishes of tropical waters. Symp. zool. Soc. Lond. 44:219-241.

Lundqvist, H., 1980. Influence of photoperiod on growth in Baltic salmon parr (*Salmo salar* L.) with special reference to the effect of precocious sexual maturation. Can. J. Zool. 58:940-944.

Macek, K.J., 1968. Reproduction in brook trout (*Salvelinus fontinalis*) fed sublethal concentrations of DDT. J. Fish. Res. Board Can. 25:1787-1796.

Maddock, B.G., 1974. A technique to prolong the incubation period of brown trout ova. Prog. Fish Cult. 36:219-221.

Magri, M.H., 1983. Action facilitante de l'hormone gonadtrope ou de steröides sexuels sur le déclenchement de la puberté chez la truit arc-en-ciel. Interférence avec la photopériode. Thèse Doct. 3éCycle, Univ. Paris VI, 74 p.

Marcel, J., 1981. Contrôle de la reproduction et gestion des gamètes de quelques espèces de poissons téleostéens. Dipl. E.P.H.E. 132 P.

McQuarrie, D.W., Markert, J.R., and Vanstone, W.E., 1978. Photoperiod induced off season spawning of coho salmon (*Oncorhynchus kisutch*). Ann. Biol. anim. Bioch. Biophys. 18:1051-1058.

McQuarrie, D.W., Vanstone, W.E., and Markert, J.R., 1979. Photoperiod induced off season spawning of pink salmon (*Oncorhynchus gorbuscha*). Aquaculture 18:289-302.

McKim, J.M., 1977. Evaluation of tests with early life stages of fish for predicting long-term toxicity. J. Fish. Res. board Can. 34:1148-1154.

Melnikova, E.L., 1982. Effect of illumination on hatching of *Salmo salar* L. embryos. 1. Character of hatching at different illumination intensity under natural temperature conditions. In: R.V. Kazokov, Biologie et Cutlure des Samonidés, GosNiorkh, Lenengrad. pp. 126-134.

Menedez, R., 1976. Chronic effects of reduced pH on brook trout (*Salvelinus fontinalis*). J. Fish. Res. Board Can. 33:118-123.

Olson, L.E. and Marking, L.L., 1975. Toxicity of four toxicants to green eggs of salmonids. Prog. Fish Cult. 37:143-147.

Pchelovodova, D.V., 1982. Quality of progeny of the rainbow trout spawners as correlated with amount of food obtained. In: R.V. Kazokov, Les particularités biologiques des reproducteurs d'espèces de poissons à valeur économique élevée, GosNiorkh, Leningrad. pp. 76-83.

Pickering, A.D. and Christie, P., 1981. Changes in the concentrations of plasma cortisol and thyroxine during sexual maturation of the hatchery-reared brown trout, *Salmo trutta*. Gen Comp. Endocrinol. 44:487-496.

Piper, R.G., McElwain, I.B., Orme, L.E., McCraren, J.P., Fowler, L.G., and Leonar, J.R., 1982. Fish Hatchery Management. United States Dept. Interior, Fish and Wildlife Service, Washington, D.C.

Pohl, M., Schmidt, R., and Holtz, W., 1982. Manipulation of spawning activity in rainbow trout by light program. In: C.J.J. Richter and J.H.Th. Goos (Comp.), Reproductive Physiology of Fish, PUDOC, Wageningen. pp 242.

Poston, H.A. and Livingston, D.L., 1971. The effect of continuous darkness and continuous light on the functional sexual maturity of brook trout during their second reproductive cycle. Fish. Res. Bull. N.Y. 33:25-29.

Pyle, E.A., 1969. The effect of constant light or constant darkness on the growth and sexual maturity of brook trout. Fish. Res. Bull. N.Y. 31:13-19.

Runnstrom, S., 1951. The population of char, *Salmo alpinus* Linné, in a regulated lake. Rep. Inst. Freshwat. Drottningholm 36:66-78.

The Washington Sea Grant Program takes pleasure in sending you this book for your review.

Title: Salmonid Reproduction: an International Symposium

Editors: Robert N. Iwamoto
Stacia Sower

Publisher: Washington Sea Grant

Publication Date: July 1985

Price: $10.00* (U.S. Funds)

ISBN: 0-934539-00-6

LC: 85-13764

CIP: yes

Distributor: Washington Sea Grant Communications
3716 Brooklyn Avenue NE
Seattle, WA 98105

Please send two tearsheets of your review upon publication to:

Laura Mason
Promotion & Distribution
Washington Sea Grant
3716 Brooklyn Avenue NE
Seattle, WA 98105

*Wash. State residents please add 7.9% sales tax

Scott, D.P., 1962. Effect of food quantity on fecundity of rainbow trout, *Salmo gairdneri*. J. Fish. Res. Board Can. 19:715-731.

Scott, D.P., 1979. Environmental timing and the control of reproduction in teleost fish. Symp. zool. Soc. Lond. 44:105-132.

Shiraishi, Y. and Fukuda, Y., 1966. The relation between the daylength and the maturation in four species of salmonid fish. Bull. Freshwater. Fish. Res. Lab., Tokyo, 16:103-111.

Skarphédinsson, Scott A.P. and Bye, V.J., 1982. Long photoperiods stimulate gonad development in rainbow trout. In: C.J.J. Richter and J.H.Th. Goos (Comp.), Reproductive Physiology of Fish, PUDOC, Wageningen. p. 243.

Smirnov, A.I., 1955. The effect of mechanical agitation at different periods of development on eggs of autumn chum salmon (*Oncorhynchus keta* Imfrasp. Autumnalis Berg. Salmonidae). Doklady Akad. Nauk SSSR 105:873-876.

Sykora, J.L., Smith, E.J., Synak, M., and Shapiro, M.A., 1975. Some observations on spawning of brook trout (*Salvelinus fontinalis*, Mitchill) in lime neutralized iron hydroxide suspensions. Water Res. 9:451-458.

Takashima, F. and Yamada, Y., 1984. Control of maturation in masu salmon by manipulation of photoperiod. Int. Symp. Salmonid Reproduction, Seattle.

Titarev, F., 1975. Acceleration of maturation in the rainbow trout (*Salmo gairdneri*) under the influence of increased water temperature. J. Ichthyol. 15:507-509.

Torrissen, O. and Torrissen, K.R., 1981. The effects of light on the mortality of different pigmented Atlantic salmon (*Salmo salar*) eggs. Inter. Council for the Exploitation of the Sea. CM 1981/M:16.

Vincent, R.E., 1980. Some influences of domestication upon three stocks of brook trout (*Salvelinus fontinalis* Mitchill). Trans. Am. Fish. Soc. 89:35-52.

Wales, J.H., 1979. Induction of hepatoma in rainbow trout *Salmo gairdneri* Richardson by the egg bath technique. J. Fish. Diseases 2:563-566.

Whitehead, C., Bromage, N.R., Forster, J.R.M., and Matty, A.J., 1978. The effects of alterations in photoperiod on ovarian development and spawning time in the rainbow trout (*Salmo gairdneri*). An. Biol. anim. Bioch. Biophys. 18:1035-1043.

Winnicky, A., 1967. Embryonic development and growth of *Salmo trutta* L. and *Salmo gairdneri* Rich. in conditions unfavourable to respiration. Zool. Polonias 17:45-58.

Woodhead, A.D., 1960. Nutrition and reproductive capacity in fish. Symp. Proc. Nutr. Soc. 19:23-28.

Woodhead, A.D., 1978. Fish in studies of aging. Exp. Geront. 13:125-140.

Wootton, R.J., 1979. Energy costs of egg production and environmental determinants of fecundity in teleost fish. Symp. zool. Soc. Lond. 44:133-159.

Wootton, R.J., 1982. Environmental factors in fish reproduction. In: C.J.J. Richter and J.H.Th. Goos (Comp.), Reproductive Physiology of Fish, PUDOC Wageningen. pp. 210-219.

Wydoski, R.S. and Cooper, E.L., 1966. Matruation and fecundity of brook trout from infertile streams. J. Fish. Res. Board Can. 23:623-649.

Zawisza et Bakiel, 1970. Gonadal development, fecundity and egg survival in *Coregonus albula* L. In: C.C. Lindsey and C.S. Woods (Editors), Biology of Coregonid Fishes, Univ. of Manitoba Press. pp. 363-397.

ABSTRACTS–Environmental Aspects

The Use of Altered Seasonal and Constant Photoperiods in the Production of All-Year Round Supplies of Eggs in the Rainbow Trout

N. Bromage, J. A. K. Elliott, and J. Springate, University of Aston, Birmingham, United Kingdom

A major constraint to the more efficient commercial production of trout is the difficulty in supply of eggs. In the U.K. eggs are commonly only available from home hatcheries from November to February/March and the season can only be extended by the importation of eggs from overseas particularly the U.S.A. and Denmark. Similar difficulties imposed by seasonal supplies of eggs are also experienced by other producing countries. Ideally, eggs should be available throughout the year or at least at times prescribed by the industry. Improvements in the availability of eggs would ensure a more effective use of farm facilities and staff and guarantee an all-year round supply of table fish.

One method of altering the time of spawning of salmonids, and hence the supplies of eggs, is to manipulate the daylength to which the broodstock fish are exposed. The present paper discusses the use of both seasonally-changing and constant light regimes in the advancement and delay of the natural spawning times of different strains of rainbow trout. Up to 4 months advancement and 3 months delay in spawning can be achieved during one reproductive season in parallel groups of broodstock using relatively simple photoperiod regimes. Subsequent manipulations over successive cycles can provide spawnings during any month of the year. It is also possible to have up to three batches of eggs from the same fish over a 16-month period.

A variety of experiments are described including pilot-scale studies, and also more importantly, data from facilities on a number of commercial farms where many hundreds of broodstock are under photoperiod control. The effects of altered photoperiod on the time of spawning and fecundity and the quality of the egg are discussed.

Control of Maturation in Masu Salmon by Manipulation of Photoperiod

Fumio Takashima and Yoshiaki Yamada, Tokyo University of Fisheries, Tokyo, Japan

In order to change the spawning time in a landlocked variety of masu salmon, *Oncorhynchus masou*, by manipulation of photoperiod, the effective light regime was investigated.

O. masou usually spawn in October at the age of two under natural conditions. However, maturation and spawning were accelerated two to three months earlier when reared under long photoperiods (18L-6D or 24 L-0D) for one or two months during the initial course of the reproductive cycle, and under a short photoperiod (6L-18D) during successive periods of the cycle. The maturation, on the other hand, was delayed when the brood fish were subjected to continuous long photoperiods. Long photoperiods during the initial course of the cycle seemed to stimulate early process in the accumulation of fatty yolk. At least, one month of a continuous long photoperiod is necessary to initiate maturation. Rapid vitellogenesis, as in the other salmonid species, was apparently stimulated by successive short photoperiods.

The fecundity of matured females under an artificial photoperiod was significantly higher than that of naturally matured ones. Although the size of the eggs was smaller, percentage of eyed eggs did not differ from that of naturally matured eggs. Serum gonadotropin did not change during a long photoperiod and increased at the time of ovulation that was triggered after transferring to a short photoperiod.

Photoperiod Control of Spawning Time in Rainbow Trout *(Salmo gairdneri)* Using Constant Long Daylengths

A. P. Scott, O. Skarphedinsson, S. M. Baynes, and V. J. Bye, Ministry of Agriculture, Fisheries and Food, Lowestoft, United Kingdom

Long daylengths of 18-20 hours have been shown to stimulate precocious maturation in male underyearling trout and to induce a six-month reproductive cycle in adults. Two trials, based upon the latter observation, have been carried out on a commercial fish farm in order to obtain out-of-season production of gametes. Both have been successful, in one case yielding gametes in July/August and in the other April/May.

The system we have used is cheap and simple to run, requiring minimal equipment and supervision. Lightproof facilities are not required. Light bulbs suspended above the broodstock tanks and controlled by time clocks are switched on from 0400 to 2200 hours every day, starting at or about the time of spawning.

One consequence of the constant long daylengths is that the growth of the broodstock is reduced. The reasons for this are discussed (with supporting experimental data). The mechanism of action of long daylengths on the reproductive cycle of the rainbow trout is also discussed.

The Effect of Size, Age, Growth Rate and Photoperiod on Maturation in the Brook Trout *(Salvelinus fontinalis)*

Stephen D. McCormick and R. J. Naiman, Woods Hole Oceanographic Institution, Woods Hole, Massachusetts

Brook trout *(Salvelinus fontinalis)* were raised at constant temperature (10°C) under two photoperiod conditions; one cycled normally with calendar date and the other was three months delayed from the norm. Within each photoperiod two growth groups, one fast (0.246 g/d over one year) and one slow (0.046 g/d), were created by controlling the amount of food offered. Photoperiod controlled the temporal phasing of the maturation cycle. Within each photoperiod, size and/or growth rate, and not age, determined the proportion of maturing individuals during their first autumn (0 +). The largest proportion of mature individuals occurred in the delayed photoperiod, fast-growing fish. Plasma thyroxine levels were generally lower in slower growing fish and showed an annual, photoperiod controlled pattern in all groups. In each experimental group, the proportion of 0 + mature females was always significantly ($p<0.05$) lower than the proportion of mature males. All fish in each experimental group became mature in their second autumn (1 +). A size related threshold which is sexually dimorphic and which interacts with photoperiod can best explain the 0 + and 1 + maturation of brook trout under artificial culture.

Photoperiod Induced Delayed Spawning of Freshwater Reared Chinook Salmon

W. S. Johnson, Ministry for Conservation, Fisheries & Wildlife Division, Alexandra, Victoria, Australia

Chinook salmon have been cultured at the Snobs Creek Freshwater Fisheries Research Station and Hatchery, Victoria, Australia, for more than 14 years. The salmon live their entire life cycle at the station in freshwater. Breeding has been relatively successful and two lakes are regularly stocked with salmon which grow quickly to a large size and provide excellent angling. In 1980 a five year program was initiated to identify factors adversely affecting breeding success and to make improvements where possible to fish husbandry.

The effect of relatively high water temperature during the maturation of adults and the incubation of ova has been studied, and related to embryonic deformation and death. An artificial photoperiod regime has been used in an effort to delay adult maturation to later in the autumn when water temperatures are considered to be more favourable.

Delayed Gametogenesis and Spawning in Rainbow Trout *Salmo gairdneri* Kept Under Permanent Light During the First and Second Reproductive Cycles

A. Bourlier and R. Billard, Laboratoire de Physiologie des Poissons, I.N.R.A., Rennes Cedex, France

Male and female rainbow trout in their first reproductive cycle were submitted on June 21, 1977 to long days (16L:8D; 24L:0D and 16L:8D), shifting progressively to 24L:0D within 6 months. The control group was kept under natural photoperiod. All the fish were reared under natural temperature fluctuating between 3 and 16°C. Mean ovulation and spermiation time was 2 months later in the long-day group than in the control group, and ovulation and onset of spermiation were spread over a long period. The quantity and quality of the sexual products obtained were comparable in all groups. During the subsequent reproductive cycle the experiment continued after the female groups were reorganized in June 1978: all females initially kept under long day (LD) were pooled; one group ($LD\text{-}24L_2$) was submitted to permanent light until March 1979 and another ($LD_1\text{-}C_2$) was put under natural photoperiod, while one part of the initial control was put under permanent light ($C_1\text{-}24L_2$); the rest of the fish remained under natural photoperiod ($C_1\text{-}C_2$). 50% of the females kept for the first time under permanent light ($C_1\text{-}24L_2$) ovulated or were about to ovulate 1 to 2 months later than the controls ($C_1\text{-}C_2$). All females initially kept in LD during the first cycle also showed delayed ovulation but only a part of them ovulated: 10 out of 12 in $LD_1\text{-}24L_2$ and 6 out of 11 in $LD_1\text{-}C_2$. In the ovary of ovulated and non-ovulated females, small vitellogenetic oocytes of various sizes were found, indicating abnormalities in the recruitment of these oocytes. The fish submitted for the first time to long days showed a delay in spawning but had a normal ovulation. However, this treatment impaired the subsequent cycle and would thus be difficult to use routinely on fish farms.

Fertilization Success and Sperm Motility of Atlantic Salmon *(Salmo salar)* L. in Acidified Waters

P. G. Daye, Daye Atlantic Salmon Corporation, Armdale, Nova Scotia; and B. D. Glebe, North American Research Center, New Brunswick, Canada

Fertilization success of Atlantic Salmon *(Salmo salar)* eggs by large sea-run males and precocious male parr in acidic waters was determined. Values of pH for milt, ovarian fluids, acidic media during fertilization, and acidic media during water hardening are presented. Duration of spermatozoa motility in acidic media is also given and discussed in relation to fertilization success. There was no effect on fertilization success at pH 5.0 or above. There was a steady decline in fertilization success from pH 5.0 to 4.0. No eggs were fertilized below pH 4.0. The pH LL50 was 4.5 for fertilization success. Duration of spermatozoa motility constantly declined at all tested levels until about pH 4.5 where the decline became extremely rapid with zero seconds of motility occurring near pH 4.0. The duration of spermatozoa motility of precocious male parr was longer than that for the large sea-run males at levels above pH 4.4. The significance of all results is discussed in respect to the recruitment of young into fish populations in acid stressed waters and acid precipitation.

Effects of Low pH on the Reproduction of Rainbow Trout *(Salmo gairdneri)*

G. S. Weiner, C. B. Schreck, and H. W. Li, Oregon Cooperative Fishery Research Unit, Oregon State University, Corvallis

Surface water acidification, presumably due to acidic precipitation, has been correlated with the decline of fish populations in regions of Europe and North America. Field studies indicate that failure in recruitment of early year-classes is an important contributor to the gradual demise of fish populations in impacted waters. Such recruitment failure may be due, in part, to disruption of adult reproductive physiology and mortality of early life history stages. Salmonid fishes are generally very sensitive to environmental acidification. We address the potential effects of acidification on the reproduction of salmonids, using the rainbow trout *(Salmo gairdneri)* as a model.

Male and female adult rainbow trout were exposed to pH levels 4.5, 5.0, 5.5, and control over the final six weeks of reproductive maturation. Fertilization was then performed within each treatment group (with rearing of offspring at the respective pH level) and between treatment fish and controls (with rearing in control water). Crosses between control fish were reared in groups at each pH level. Offspring were monitored for rates of survival to eye-up, hatching, and absorption of the yolk-sac.

Analysis of the results of the cross-breeding matrix yields a comparison of the effects of low pH on gametogenesis with effects on developing embryos and larvae. Plasma samples collected from adults prior to acid exposure, during the exposure, and at spawning are being analyzed for concentrations of reproductive hormones and electrolytes as indicators of effects on reproductive physiology. Indicators of effects on egg and sperm quality are also being evaluated.

Maturation Success of Pink and Coho Salmon Held Under Three Salinity Regimes

Alex C. Wertheimer, Auke Bay Laboratory, National Marine Fisheries Service, Auke Bay, Alaska

Adult pink and coho salmon were held in freshwater, seawater (28–32 ppt), and estuarine (24–31 ppt, with a lower salinity surface lens) environments to determine the effects of these salinity regimes on survival to spawning and gamete viability. The salmon were captured at first entry to freshwater or in the estuary and held until they were judged fully mature or had died. Gamete viability was determined by individual matings between and within holding environments. Blood and gonadal fluid samples were taken at spawning for osmolality and potassium analysis.

Maturation success, in terms of both survival to spawning and gamete viability, was significantly reduced for both species of fish held in seawater relative to those held in freshwater. The presence of the low salinity lens in the estuarine holding environment eliminated or moderated the detrimental effects of seawater maturation. For pink salmon, there was no difference in survival or gamete viability between fish held in freshwater and estuarine environments. For coho salmon, fish held in the estuarine and seawater environments had significantly lower survival and gamete viability than those held in freshwater, but fish held in the estuarine environment had significantly higher survival to spawning than those held in seawater. Both maternal and paternal effects contributed to reduced gamete viability.

Higher average levels of blood and gonadal fluid osmolality and potassium were associated with the groups of brood fish that had lower average gamete viability. However, correlations between gamete viability of individual matings and osmolality or potassium levels of the parents were poor.

Temperature Effect on Egg Viability and Spawning Time of Cutthroat Trout

Robert G. Piper, U.S. Fish & Wildlife Service, Bozeman, Montana

Westslope cutthroat trout egg viability and spawning time were compared in colder, fluctuating temperature creek water and constant 50°F temperature spring water.

Eggs from broodfish held in creek water were easily expelled and normal appearing. Eggs from broodfish held in spring water, however, were usually difficult to expell, and broken and opaque eggs were common. Egg viability was greatly affected by water temperature; there was a 77% eye-up in eggs from broodfish held in creek water, while only a 25% eye-up occurred in eggs from broodfish held in constant temperature spring water.

Duration of spawning varied; groups of broodfish held in creek water spawned over a 20 to 26 day period, while 44 to 57 days elapsed before all groups of broodfish held in spring water had spawned.

Effects of Temperature and Food Availability on Maturation of Sockeye Salmon in the Ocean

T. Nishiyama, University of Alaska, Fairbanks; and T. Minoda and T. Fujii, Hokkaido University, Hakodate, Japan

The return timing of salmon to fresh water is believed to result from the effects of oceanographic conditions during the last marine life stage. Specifically it has been found that the return timing of Kvichak river sockeye is inversely related to the mean sea surface temperature in the central Bering Sea during June, a month prior to the return migration. Nishiyama (1982) offered an explanation of the mechanisms determining the return timing. First, the return timing is controlled by the maturity condition of salmon which is directly regulated by sea temperature. Second, the temperature influences upon food availability and growth must affect the maturity condition and thereby the return timing.

To examine these mechanisms, the relationships between return timing, temperature and zooplankton biomass in the basin and shelf areas of the Bering Sea were investigated based on data collected during June and July from 1957 to 1981.

The zooplankton biomass was low from 1957 to 1963, but high between 1965 and 1969. Since 1974 the zooplankton biomass has tended to decrease slightly. Mean zooplankton biomass has generally been 25% higher in the shelf area than in the basin area. In the basin area, the zooplankton biomass has tended to decrease linearly with an increase in temperature, indicating that the zooplankton production was higher in the colder years. In the shelf area, the zooplankton biomass was low both in the colder and warmer years, but high in the moderate years. The peak return date of Kvichak river sockeye appeared to be related to the zooplankton biomass in the Bering Sea. A high zooplankton biomass in the basin area was associated with delayed return timing; whereas, in the shelf area, the inverse was true. Finally, the effects of temperature and food availability are discussed in relation to the vertical distribution patterns of sockeye and zooplankton.

Artificial Hatching Substrate: Effect on Yolk Sac Absorption and Growth Rate During Startfeeding of Atlantic Salmon *(Salmo salar)*

Tom Hansen and Dag Møller, Institute of Marine Research, Direktorate of Fisheries, Matredal, Norway

The Norwegian fish farming industry experiences serious mortality problems both in the hatcheries and during the startfeeding period. Both natural and artificial substrates have been reported to increase fry size and to lower mortality rate. The effect of these substrates has however not been studied in detail. The purpose of this experiment was to investigate to what extent an artificial substrate affects the growth rate, survival rate and yolk conversion efficiency to the fry.

In this experiment groups of pooled Atlantic salmon *(Salmo salar)* eggs were hatched in a California hatching system with and without an Astro-turf artificial substrate. At 41 days after hatching, each group was transferred to separate feeding units and was fed dry feed pellets for a period of 46 days. Alevins reared in Astro-turf absorbed their yolk sac faster and more efficiently than alevins reared on a flat screen. The alevins reared in Astro-turf showed no constrictions in their yolk sacs, a feature which was notable in the yolk sacs of the alevins reared on a flat screen. The fry hatched without Astro-turf grew faster than the fry hatched with Astro-turf during the first startfeeding period. However, the growth rate during the first startfeeding period is highly influenced by the amount of yolk left in the yolk sac. In this experiment the early growth of the fry hatched with Astro-turf probably could have been improved by introducing food earlier. From startfeeding day 18 and to the end of the startfeeding period the fry hatched with Astro-turf grew better than fry hatched without Astro-turf. The mortality of alevins and fry hatched with Astro-turf was lower both in the hatchery and during startfeeding.

Effects of Sediment on the Reproduction of Salmonids

K V. Koski, Auke Bay Laboratory, National Marine Fisheries Service, Auke Bay, Alaska

Salmon and trout have evolved similar strategies for reproduction. The construction of redds in gravel substrates is a means of protecting their eggs from predation and ravages of the environment. Fry recruitment is dependent on the intra-gravel environment of the redd, and natural or man-induced events can cause significant alteration resulting in high mortality. A natural component of the redd is sediment (i.e., coarse sand), however, in high quantities it can significantly reduce the number of emerging fry. Sedimentation of streams can occur as a result of industrial development, logging, road-building, etc. and is one of the major factors affecting spawning habitat. Sediment causes mortality by physically blocking the emergence of fry from the gravel or by reducing the amount of dissolved oxygen in the intergravel water by preventing interchange with surface water. Sediment content of the spawning gravel sets the upper limit on survival, and other factors, such as dissolved oxygen, freezing, scouring, etc., impose additional mortality. The stage of development, condition, and timing of emerging fry are altered by increases in sediment in the redd. Egg size may form the basis of preferred spawning habitat by the different species of salmonids.

Alteration of spawning habitat is likely to continue and improved methods of evaluating impacts are essential. Enhancement of spawning habitat must address substrate composition. Research during the last 20 years in Alaska, Oregon and Washington has resulted in equations expressing the relationship between sediment and survival to emergence of salmonids. New methods of gravel analysis have been developed. New expressions of gravel composition and survival are developed, enabling one to predict survival to emergence and to determine quality of spawning habitat.

Salmonid Redd Dewatering: What Do We Know?

C. D. Becker and D. A. Neitzel, Battelle, Pacific Northwest Laboratory, Richland, Washington

Spawning areas used by adult salmonids may be dewatered by environmental manipulation during the critical period when embryonic developmental phases are intergravel. As a result, eggs and yolk-sac alevins may be exposed to altered physical conditions imposed by withdrawal of water from the gravel. Dewatering may be periodic and short-term, as imposed by daily power peaking operations of a hydroelectric generating facility, or sporadic and long-term, as imposed by drawdowns for irrigation or construction activities.

We have conducted dewatering experiments with chinook salmon and rainbow trout eggs in artificial redds at our laboratory since 1979. The data indicate that, relatively, egg phases are tolerant of dewatering while alevin phases are intolerant. Comparative quantitative data are available. A number of physical factors that influence survival of intergravel development phases during dewatering were examined experimentally. Other factors that might influence survival of development phases were identified from the literature. Many of these factors, such as subsistent flow from bank storage after drawdown, will vary with seasonal and site-specific conditions. Knowledge of relative tolerance of intergravel phases to dewatering, and of the effect of various abiotic conditions associated with dewatering, can be used in developing and implementing mitigation measures.

Integration of Salmon Enhancement in Water Quality Planning: A Case Study

Thomas B. Murdoch, Snohomish County Public Works Department, Everett, WA

Snohomish County, Washington, is a rapidly urbanizing area in western Washington that contains over 3,000 miles of streams and 600 lakes. Most of these bodies of water support salmonids; however, increased development without proper drainage management is destroying spawning and rearing habitats at an alarming rate.

To further public appreciation for watershed management programs that avert stormwater damage to property and also protect natural resources, the Snohomish County Public Works Department has sponsored stream enhancement projects for adoption by community groups. Such organizations as schools, sportsmen's clubs, ad-hoc committees and civic organizations have reared salmon in egg boxes, replanted streambanks, removed debris blocking stream channels, and identified unmapped small streams.

Results are provided to officials making land use decisions. Of particular importance is the identification of formerly unknown streams and the incorporation of these data into local zoning maps. Streams "adopted" by a neighborhood group have an improved chance for survival. When organizations become actively involved in local water quality projects, dramatic positive results can be achieved in short periods, with salmon being the primary beneficiary.

Environment and Genetics Shapes Evolution?

Abe Vanderhorst, Salmon Troller, Nanaimo, British Columbia, Canada

Can it be said that Environment plus Genetics will shape the Evolution of a species and thus that the "Oregon Problem" will continue and be felt elsewhere? In the troll fishery on the West Coast of Vancouver Island a difference in size between marked hatchery coho and unmarked and marked "wild coho" was noted.

For at least four years running (1979 to 1982 inclusive) adipose fin clipped coho salmon were always found to be much more prevalent when fishing on small coho than when fishing on large coho regardless of time of season. Work noting the relative size of all tagged fish caught and finding their origin was done. It appeared that hatchery fish from more recently constructed hatcheries (mainly Canadian) were larger than fish from older hatcheries (mainly American).

Research and other literature were obtained to find if this had been appreciated by anyone else. Hatchery techniques of attempting to get excellent immediate returns were looked at. The creation by hatcheries of large numbers of jacks with the culling out of these jacks is suspected and questioned.

Overall with the manipulation of the salmon's early life environment by hatcheries such as the use of rapid growth feeds and temperatures, the salmon are evolving into a lesser species.

Nutrition

Session Leader
Ronald W. Hardy

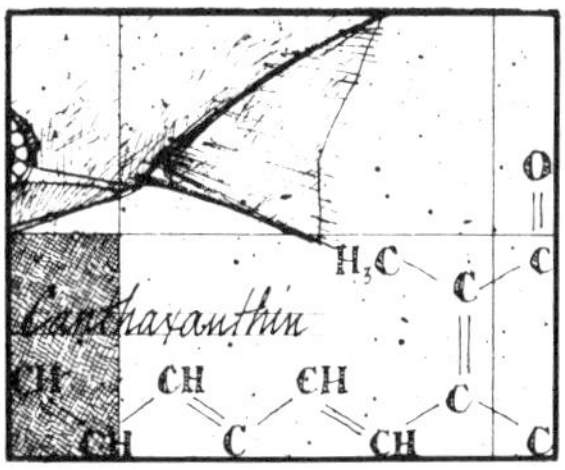

SALMONID BROODSTOCK NUTRITION

Ronald W. Hardy[1]

Abstract: Nutrition influences reproductive performance of trout and salmon by affecting fecundity, egg size, egg chemical composition, egg fertility, egg hatchability, fry viability, and pre- and post-spawning mortality. To facilitate the understanding of nutritional effects on salmonid reproduction, various nutritional aspects are discussed separately. Pertinent literature is reviewed, current knowledge in each nutritional category is summarized, and areas in which future research is needed are discussed.

Introduction

Intensive aquaculture is dependent upon constant and reliable production of viable fry to make maximum use of capital-intensive rearing facilities. Annual production of salmon and trout eggs from cultured broodstock in the United States is over 10 million and 250 million, respectively. Raising fish as broodstock is a potentially lucrative enterprise since the value of the eggs produced by each fish is about five times greater than the value of the same fish in the fish market. Because the eggs have a high economic value, prespawning mortality, postspawning mortality (in trout), inviable eggs, and poor quality fry can cause substantial losses. The extent to which the diet can increase fecundity and egg viability and decrease prespawning and postspawning mortality is at present an open question but one which merits attention. The potential for improvement is encouraging, especially when one considers the gains obtained in poultry and livestock production during the past 50 years. Cowey (1982) suggested that "considerable benefit and advances are likely to flow and, in fact, are already flowing to aquaculture from an improved understanding of the physiological basis of fish production." At present, we are at the same point in our understanding of the physiological and biochemical basis of fish production and fish reproduction as were animal scientists 30 years ago. By analogy, we can expect marked improvements to occur as worldwide research efforts continue to increase our basic understanding of fish biochemistry and physiology, and the accumulated knowledge is applied to fish husbandry and nutrition. This review will outline the current understanding of the role of nutrition in salmonid reproduction, and will suggest possibilities for future research and for practical application of results on diet formulation and feeding strategies.

In the context of its effects on reproduction, nutrition can be divided into the categories of feeding level, dietary protein/energy levels, essential nutrients (vitamins, minerals, amino acids, and essential fatty acids), and additional dietary ingredients (such as carotenoid supplements). These nutritional factors can

[1] National Marine Fisheries Service, Northwest & Alaska Fisheries Center, Seattle, WA

influence several aspects of reproduction in salmonids. In eggs, they can influence size, chemical composition, fertility, and hatchability. Also affected are fry viability and size and pre- and postspawning mortality of broodfish. This review will cover the effects of diet on female salmon and trout reproduction only. At present, there is no experimental evidence indicating that diet influences sperm quality in salmonids.

Feeding Level

Once a group of fish are identified as broodstock, conventional rules governing feeding levels no longer seem to apply. Optimum feed conversion values are no longer a primary goal. Optimum reproductive performance, which can be fecundity, percent hatchability, cost per 1000 eggs or other measure, is now the primary goal. Determining the proper feeding level for broodstock salmonids depends on several factors, including the way in which optimum reproductive performance is defined. As the research results listed below illustrate, apparent conflicting reports are often the result of using different ways of assessing optimum reproductive performance in feeding level studies. Gutsell (1940) published the first study concerning the effect of quality and quantity of diet on trout reproduction. He used trout spawning as four-year olds and varied the feeding level during the 8-month period before spawning. Gutsell found that feeding to satiation resulted in increased egg size and a greater number of eggs per fish than feeding at one half or one-fourth satiation. Fewer eggs of females fed to satiation reached the eyed stage than did eggs of females at other feeding levels. The cost per eyed egg from the trout fed to satiation was three times higher than from the group fed at one-half satiation and four time higher than from the group fed at one-fourth satiation. Feeding to satiation did not influence the quality of the eggs from trout spawning as 2- or 3-year-olds but did increase the cost. Feeding level did not influence the percentage of fish spawning.

Scott (1962) restricted food intake of rainbow trout during various periods before spawning and reported that severe food restriction resulted in a reduction in the percentage of fish spawning and in fecundity among fish that did spawn. Egg size was not affected by food intake. Scott concluded that fecundity differences between fish in dietary treatments were the result of follicular regression (atresia). Bagenal (1969), in a more complete study, reported that a higher percentage of well-fed than under-fed trout mature and that the well-fed trout contained significantly more eggs, even after adjusting for their larger body size. However, the eggs from well-fed fish were found to weigh less on a dry weight basis than the eggs from the underfed trout. Bagenal concluded that atresia could not possibly account for differences in fecundity.

Orr et al. (1982) raised two groups of rainbow trout, from 1 year of age to spawning at 2 years of age, on a hand-fed, restricted ration (about 1% body weight per day) or on an unlimited ration using demand feeders. The group of trout fed with demand feeders weighed an average of 2.8 kg compared to 1.7 kg for the group fed the restricted ration. Food conversion was slightly better for the demand-fed group. Fecundity and egg size were greater for the demand-fed group, while relative fecundity (eggs per kg body weight) was lower. Egg survival to the eyed stage was similar between groups. Although the feed cost per 1000 eyed eggs was 30% higher in the demand-fed group, the total cost of production favored feeding the broodfish *ad libitum* with demand feeders. This takes into account labor and the fact that 30% fewer fish were required to produce a given number of eggs.

Harris and Griess (1978) found that using a feeding level of 0.4% body weight per day resulted in a reduction in fecundity and egg size in rainbow trout. Trout fed 0.6 or 0.8% body weight per day had the same fecudnity and egg size. Survival of the eggs to the eyed stage was not influenced by feeding level.

Roley (1983) fed two groups of rainbow trout for one year prior to spawning at two feeding levels—satiation and one-half satiation. He observed increased total fecundity in trout fed to satiation but higher relative fecundity (number of eggs per kg body weight) in the group fed at the lower feeding rate. Egg size and percent egg survival to hatch were high in the trout fed one-half satiation. Roley found an effect of feeding level on reproductive performance independent of and in addition to the effect of feeding level on fish weight at spawning.

Ridelman et al. (1984) compared reproductive performance of rainbow trout fed *ad libitum* to the performance of fish starved for the 45-day period before spawning. There were no differences in fecundity, egg physical or chemical characteristics, or egg viability between the two groups.

From the preceding discussion, it can be concluded that feeding level influences fecundity, egg size, the percentage of eyed eggs, and the cost per egg in rainbow trout. Feeding level seems to exert its effect mainly by controlling the size of the fish at spawning. Larger fish contain more eggs, but these eggs may be smaller than those from trout fed at a lower rate. Some studies have shown that the survival to swim-up of offspring from smaller eggs is reduced. Determining the appropriate feeding level for broodstock trout will depend on the priorities of each aquaculture operation—economy, maximum egg production, or some combination of the two.

Diet Composition

Many fish culturists have noticed that some diet formulations seem to support better salmonid reproductive performance than other diet formulations. Researchers have attempted to identify specific diet formulations or dietary ingredients that influence reproductive success. Gutsell (1940) was also the first researcher to demonstrate an effect of diet composition or quality on reproductive performance of salmonids. He fed an "unbalanced" diet containing only sheep liver, wheat middlings, and meat meal and reported that a considerable number of ruptured eggs were present at spawning. Other dietary combinations resulted in reductions in fecundity and increases in egg losses during incubation.

Dumas (1961) found that brown trout fed a pellet and meat diet produced smaller eggs that differed in proximate composition from eggs spawned from brown trout fed a standard hatchery diet. Unfortunately, he did not report fecundity or percent hatch data for the two treatment groups.

Satia (1973) compared the reproductive performance of rainbow trout fed four diets and obtained the highest survival of eggs to hatching in the treatment group receiving a diet containing 25% shrimp wastes. Satia reported that egg hatchability was not affected by the weight or size of the egg but by the quantity and quality of protein, lipid, and ash in the eggs, which differed between dietary treatments. Iodine number of the egg lipids was positively correlated with egg hatchability. Satia concluded that shrimp waste contained unidentified factors responsible for increased egg production and hatchability of rainbow trout eggs. Trout fed a commercial trout diet had the poorest egg quality compared to the other diets used in Satia's experiments.

Ridelman (1981), using the same stock of fish, could not find any relationship between the proximate composition of the trout eggs and their hatchability. Other authors have been unable to correlate the hatching success of trout or salmon eggs with egg proximate composition (Watanabe et al., 1984).

What these researchers probably observed in their experiments were the results of feeding diets that were deficient in some essential nutrient, or were otherwise imbalanced. In order to support maximum reproductive performance of broodstock salmonids, it is desirable and economical to identify dietary requirements and formulate accordingly.

Protein/Energy Levels

It has been a continuing belief among fisheries scientists that the dietary requirements of a reproducing salmonid will be higher than those of a growing but nonreproducing fish. There are several reasons for this belief. Animal and poultry scientists have found that certain nutrients are needed in higher amounts during egg laying or pregnancy. In nature, trout and salmon feed heavily on high-protein, high-fat materials before spawning and often do not eat during the final stage of ovarian development. Further, it has been observed that in captivity fish (especially salmon) generally produce eggs whose quality is inferior to those of wild fish. Protein and energy levels are frequently targeted as critical dietary components that should be increased in broodstock salmonid diets. It is thought that the dietary protein requirement is higher during ovarian development because the fish needs to synthesize vitellogenin for deposition in the eggs. Higher dietary energy levels are presumably needed to supply the metabolic energy necessary for synthesizing the material that fills the eggs. The problem with these beliefs is that somatic growth essentially stops during ovarian development. In addition, the proximate composition of ovarian tissue is similar to fish somatic tissue so that fish may simply switch from synthesizing fish tissue to synthesizing ovarian tissue, and this may not require an increase in metabolic effort. The experimental evidence supporting these beliefs is far from conclusive but, in some situations, indicates that dietary protein and energy levels affect reproductive performance.

Several experimenters have attempted to determine the protein and energy requirements of broodstock trout. The first efforts were those of Phillips et al. (1964), who fed three diets containing 43%, 37%, and 31% protein and 2900, 2550, and 2050 Kcal ME/kg, respectively, to brown trout broodstock. The feeding level of the low-protein, low-calorie group of fish was slightly higher than the other two dietary treatments and, at spawning, the average weights of fish in the three groups were nearly the same. Fecundity was slightly higher among fish fed the high-protein, high-calorie diet. Egg size and percent hatch were inversely related to dietary protein/calorie level. However, the diets of Phillips et al. (1964) differed not only in protein and energy levels but also in ingredients and their levels. Although the results of Phillips et al. showed an effect of diet on reproductive performance, one cannot conclude that the observed differences were only due to differences in dietary protein and energy levels.

Smith et al. (1979) conducted a 2½ year feeding trial in which rainbow trout broodstock were fed diets ranging from 36% to 48% crude protein and 2500 to 3450 Kcal D.E./kg, respectively. During the study, the trout spawned three times. While the weight of the fish and total fecundity significantly increased with increasing dietary protein level, the percent eyed eggs of each year's spawn was not influenced by diet. With increasing fish age, however, percent eyed eggs decreased. Although the low-protein diet was the most expensive per kg weight gain, the feed cost per 1000 eggs was nearly identical for the different diets. Pre- and postspawning mortality was nearly the same among dietary treatment groups. The diets of Smith et al. (1979) differed from one another in ingredient composition, but the differences were not as great as in the diets of Phillips et al. (1964).

Takeuchi et al. (1981) reported high hatchability of eggs from rainbow trout broodfish fed a low-protein, high-calorie diet (36% protein and 18% lipid). A subsequent experiment (Watanabe et al., 1984) showed that a reduction of the protein level to 28% resulted in decreased growth and a slight reduction in egg size and percent hatch from 86% to 70%. The diets of Watanabe et al. differed in the percent of fish meal, wheat flour, feed oil.

Roley (1983) formulated four isocalorie diets ranging from 27% to 56% protein which contained 3800 Kcal ME/kg diet. These diets were fed to rainbow trout broodstock for 8 months before spawning which occurred at 2 years of age. Roley found that fish fed diets containing 27% and 37% protein weighed significantly less at spawning than fish fed diets containing 47% and 56% protein. Total fecundity and egg diameter were highest in fish fed the 47% protein diet, but no statistically significant differences between treatments were found. Survival to hatching was similar among dietary treatment groups. Roley concluded that the dietary protein requirement for optimum reproductive performance was between 37% and 47% in a diet containing about 3800 Kcal ME/kg.

From these studies, one can conclude that protein and energy levels in broodstock diets influence reproduction in trout, primarily through their effects on fish weight at spawning. The dietary protein requirement of trout yearlings is thought to be 35% (National Research Council, 1981). However, Roley found that in a broodstock diet 37% protein was not as good as 47% perhaps because he used fish which spawned at 2 years of age. Smith et al. (1979) pointed out that although high-protein, high-calorie diets are more expensive than diets lower in protein and calories, improved feed conversion and decreased food intake offset the cost. They suggest feeding diets high in protein and energy for the first 2 or 3 years to obtain maximum size during the period of intensive somatic growth. Then a switch may be made to diets lower in protein and energy. The actual optimum dietary protein and energy levels for broodstock salmonids probably vary depending on age at spawning, stock and species of fish, dietary ingredients, and other factors. Further research is needed before a concrete recommendation can be made.

Specific Essential Nutrients

Relatively few qualitative and quantitative studies with specific nutrients, such as amino acids, vitamins, minerals, and essential fatty acids, have been conducted with broodstock salmonids. There are several reasons for this. First, to conduct quantitative studies with specific nutrients, one must often use purified diets. Yu et al. (1979) only recently demonstrated that trout could be raised successfully from fry to spawning, a period of 34 months, on a purified diet. The cost of the purified diet is currently about $12 U.S. per kg but it can be substantially higher if specialized vitamin premixes or highly purified oil sources are used.

A second factor that may complicate studies of broodstock nutrient requirements is that the nutritional history of the fish is critical. The fish must be raised on special diets for literally years to avoid confounding the study with carryover of nutrients stored in the body. Finally, there is a lack of non-destructive microchemical assays to determine nutrient status. It is difficult to justify sacrificing valuable broodfish to obtain tissues for chemical analysis. It takes a substantial commitment of money, time, and resources to carry out any broodstock study, especially when the results are apt to be less than spectacular. Despite these pitfalls, some progress has been made in the areas of dietary vitamin, mineral, and fatty acid requirements. No studies have been conducted on amino acid requirements of broodstock salmonids, but changes in their dietary need should be less likely than changes in other categories of nutrients during ovarian development.

Vitamins

Information on the effects of dietary vitamin levels on egg vitamin levels and egg hatchability has begun to accumulate within the last decade. Kinumaki et al. (1972) conducted feeding trials with rainbow trout broodstock to study the relationships between dietary fat soluble vitamin levels, the levels of these vitamins in maternal tissues and eggs, and hatchability of the egg.

Diets containing basal and supplemented levels of vitamins A, D, and E were fed to rainbow trout at 1% body weight for 3½ to 4 months before spawning. Increased dietary levels of vitamin A and vitamin E resulted in increased levels of these vitamins in the eggs at spawning. Egg vitamin D levels were not substantially increased in fish receiving diets supplemented with vitamin D above the control diet level (46 I.U./100 g diet). High dietary levels of vitamin E appeared to increase egg vitamin A levels, but high dietary levels of vitamin A appeared to decrease egg vitamin E levels. Vitamin E was withdrawn from maternal tissues and transferred to the developing ovaries in fish fed diets containing 3 mg vitamin E per 100 g diet, but maternal stores remained high in fish fed diets containing over 10 mg vitamin E per 100 g diet. Egg size was not affected by dietary fat-soluble vitamin level. Percent eyed eggs was slightly reduced in treatment groups fed diets containing high dietary levels of vitamin A and D in one experiment, and vitamin E in another.

The relationship between dietary and egg ascorbic acid levels have been investigated by several researchers. Ascorbic acid is of particular interest in broodstock diets because of the high levels found in salmonid eggs and because ascorbic acid is easily destroyed during manufacture and storage of pelleted feeds. Hilton et al. (1979) examined the ascorbic acid concentration of a number of tissues and organs in rainbow trout and found that ovary levels were among the highest. Underyearling, yearling, and mature rainbow trout had levels of 326, 548, and 451 μg ascorbic acid per g wet ovary, respectively. Dietary levels of ascorbic acid were 154, 127, and 116 ppm for the three groups. Increasing the dietary ascorbic acid supplementation from 1045 to 10450 ppm in a dry diet formulation (430 and 2200 ppm were present after pelleting and storage) did not affect coho salmon egg hatchability, but the ascorbic acid level of the eggs was increased from 310 to 510 μg/g wet weight (Hardy, unpublished data). Fry viability was not assessed.

Ridelman (1981) compared egg ascorbic acid levels between wild steelhead and rainbow trout fed diets containing either 800 ppm or 1400 ppm ascorbic acid. The egg ascorbic acid levels were 523, 354, and 504 μg/g wet tissue respectively. There were fewer "blank" lots of eggs from trout fed the higher level of ascorbic acid. Sandnes et al. (1984) compared egg hatchability of rainbow trout fed either an ascorbic acid-supplemented diet (115 mg/kg diet) or a nonsupplemented diet. These diets were fed for 3 months prior to spawning. Egg ascorbic acid levels of 15 μg/g wet weight in the unsupplemented group and 31 μg/g wet weight in the supplemented group were reported, well below the levels reported by other workers. Egg hatchability was reduced from 87% in the supplemented group to 62% in the unsupplemented group. Sandnes et al. (1984) suggested that broodstock trout diets should be supplemented with enough ascorbic acid to insure a level of at least 100 ppm at feeding.

The studies reported above clearly show that dietary vitamin levels can influence the levels of vitamins in salmon and trout eggs. Hatchability can be increased when dietary intake is raised from deficient to adequate levels, but additional supplementation does not necessarily increase hatchability further and may, in the case of fat-soluble vitamins, actually decrease egg hatchability. Fry survival to swim-up and beyond must be investigated further to determine if increased egg vitamin levels can affect fry viability. In addition, the effects of increased dietary vitamin supplementation on mortality of broodfish should be examined.

Minerals

Early work in the area of minerals was confined to attempts to discover relationshps between hatchability and chemical composition of trout eggs. Hirao et al. (1955), for example, reported that the content of iron in the eggs was proportional to the hatching rate, an observation quoted by Love (1970). Using rainbow trout, Takeuchi et al. (1981) were the first demonstrate the effect of maternal dietary trace mineral deficiency on on egg quality. Deficiencies were produced by deleting the trace mineral premix from the diet for 2½ years prior to spawning. In the fish fed the deficient diet, total egg production and female weight at spawning were reduced. Egg diameter was not affected, but percent survival to hatch was 0.4% compared to 87.2% in the control group. The concentrations of manganese, zinc, and iron were significantly reduced in the maternal trout bones while only manganese was significantly reduced in the eggs. Hardy et al. (1984), working with coho salmon broodstock, conducted feeding trials to determine if there was a need for additional trace mineral supplementation above the dietary levels used in production diets for fingerling salmon in order to produce high quality eggs and maintain maternal stores during ovarian development. After supplementation, they could find no evidence of maternal depletion of higher levels or trace minerals in the eggs.

Other attempts to correlate hatchability with egg chemical composition using Atlantic salmon eggs obtained from various locations have been made (H.G. Ketola, personal communication, 1983). Reduced zinc levels were found in eggs having reduced hatchability. Atlantic salmon broodstock fed diets fortified with zinc, manganese, and iron for three months before spawning produced eggs with 51% hatchability. This was nearly identical to the hatchability (52%) of eggs from broodfish fed nonfortified diets. Hardy et al. (1984) observed that the total amount of zinc or manganese deposited in the ovaries of coho salmon during the last three months of ovarian development amounted to less than 5% of the total zinc or manganese in the body of the female salmon. Based on this observation, it is not surprising that broodfish used in short-term studies showed no benefit from trace mineral fortification.

Essential Fatty Acids

Two approaches have been used to determine if dietary fatty acid levels influence egg quality of trout and salmon raised in hatcheries to spawning. One approach is to analyze the fatty acid profile of eggs from hatchery fish and to compare the results to the fatty acid profile of wild fish. The underlying assumption is that the fatty acid composition of eggs from wild fish is desirable while that of hatchery fish is a reflection of dietary levels, which may contribute to reduced egg hatchability. The other approach is to feed different dietary levels of fatty acids and measure the effects on egg fatty acid composition and hatchability.

Yu et al. (1979) fed rainbow trout purified diets containing 1% ethyl linolenate plus 1% ethyl linoleate or 1.5% ethyl linoleate for 34 months, from the fry stage (0.4 g) to spawning. The fish matured and produced viable eggs that did not differ in percent fertility or hatchability from those of fish fed a control diet (Oregon Moist Pellet). Egg size and fecundity were equivalent between dietary groups. After hatching, the fry were fed for three months to assess the effect of maternal diet on fry growth performance. Second generation fry from both dietary treatment groups grew normally without excessive mortality. The results indicate that n-3 fatty acids support normal reproduction and that n-6 fatty acids are unnecessary in the diet. The fatty acid composition of the egg phospholipids was influenced by diet.

Fish that had received linoleic acid in the diet produced eggs containing 20.4% n-6 fatty acids in the phospholipids. The eggs of fish whose diet did not contain linoleic acid had 0.5% n-6 fatty acids in the phospholipids. The n-6 fatty acids that differed in percentage between dietary treatment groups were 18:2 n-6, 20:2 n-6, and 20:4 n-6. Others have compared the fatty acid profiles of hatchery and wild Atlantic salmon and found that wild fish eggs had lower levels of 18:2 n-6 and higher levels of 22:0, 22:1 and 24:1 (H.G. Ketola, personal communication, 1983).

Hardy (unpublished data) compared the fatty acid composition of neutral and polar lipids from eggs of wild and pen-reared coho salmon and found the wild fish eggs had higher levels of 20:5 n-3 and 22:5 n-3 and lower levels of 18:2 n-6. The levels of n-6 and n-3 fatty acids were 2.2% and 36.5% in neutral lipids of eggs from wild fish and 10.4% and 22.5% in eggs from pen-reared fish. Polar egg lipids from wild fish had 1.2% n-6 and 48.2% n-3, while those from pen-reared fish fed pelleted diets had 4.1% n-6, and 36.5% n-3. Feeding a greater level of fish oil (4% or more) in the diets of the pen-reared coho salmon did not change the fatty acid composition of the eggs.

Rainbow trout fed a diet deficient in n-3 fatty acids for 3 months before spawning showed a reduction in egg hatchability compared to trout fed a diet supplemented with essential fatty acids (T. Watanabe, personal communication, 1983). Compared to the control group, eggs from fish fed the diet difficient in essential fatty acids had higher levels of 18:2 n-6 and 20:4 n-6 and lower levels of 20:5 n-3, 22:5 n-3, and 22:6 n-3 in both the triglyceride and polar lipid fractions. However, eggs from another dietary treatment group were also low in n-3 fatty acids, yet egg hatchability was not reduced. The relationship between egg fatty acid composition and egg viability is complicated, and more studies must be conducted before dietary guidelines can be established. However, the studies listed above show that dietary fatty acid intake can influence egg fatty acid composition and egg hatchability in trout.

Dietary Carotenoid Pigments

Several studies have been conducted to examine the need for dietary carotenoid pigments in salmonid broodstock diets. The results of some of these studies seem contradictory. Deufel (1965) fed groups of fish a diet containing 40 ppm canthaxanthin and he observed an increase in the percentage of fish maturing. Egg fertility was slightly increased, from 96% to 99%. Reinitz et al. (1977) fed dry diets containing 25% shrimp meal to rainbow trout broodstock for three months before spawning but could not demonstrate any effect on spawning or egg quality. Quantz (1980) could not demonstrate any effect of dietary supplementation with canthaxanthin or astaxanthin on rainbow trout egg fertility. In his study the fish were fed 8–15 weeks prior to spawning. Morrison and Smith (1981) fed diets supplemented with 55 ppm canthaxanthin or 12 ppm astaxanthin to brook trout for 66 days prior to spawning and found an increase in percent eyed eggs compared to a control diet-fed group. This increase was not statistically significant, however, and the authors recommended a longer feeding period be used to evaluate the effects of dietary carotenoid pigments on egg quality.

Harris (1984) fed diets containing 20 and 40 ppm canthaxanthin to rainbow trout broodstock for 3- and 6-month periods before spawning. The percentage of females spawning increased from 78.6% to over 91%. Fecundity and percent eyed eggs were similar among treatment groups. The intensity of egg pigmentation was higher among eggs from fish fed the experimental diets for 6 months than it was for those fed for 3 months. Tacon (1981) speculated that carotenoid pigments might be involved in egg respiration or related to tolerance of environmental factors, such as low dissolved oxygen concentrations, elevated water temperatures and ammonia levels, and harmful effect of light. Torrissen (1984) examined the sensitivity to light of Atlantic salmon eggs and found that it increased with increasing amounts of carotenoid pigments in the eggs. There was no effect of carotenoid levels on survival of eggs or alevins in darkness.

Common sense tells us that carotenoids must play an important role in the egg during the period between spawning and first feeding, since salmonids deposit carotenoids in eggs at fairly high levels. Research results to date, however, have not demonstrated what that role might be. To demonstrate if the speculative functions mentioned by Tacon (1981) exist, experiments should be conducted which measure respiration of trout and salmon eggs containing various carotenoid pigments.

Summary

In summary, research has shown that the dietary conditions (such as underfeeding, deficiencies of certain essential nutrients, and a dietary protein level below 35%) that reduce growth in non-reproducing salmonids also reduce reproductive performance by affecting size of fish at spawning, egg quality, or both. The consensus among fisheries professionals is that the nutritional requirements of reproducing salmon and trout ought to be higher than those of growing, non-reproducing fish. Despite the fact that the specific nutrient requirements of salmonid broodfish are not known, current practical diets are sufficiently well formulated to produce acceptable reproductive performance in many situations. If continuing efforts are made to determine the effects of nutrition on salmonid reproduction, progress in formulating better broodstock diets should accelerate.

References

Bagenal, T.B., 1969. The relationship between food supply and fecundity in brown trout, *Salmo trutta* L. J. Fish. Biol. 1:167-182.

Cowey, C.B. (Editor), 1982. Introduction. Special Issue Comp. Biochem. Physiol. 73B, No. 1, p. 1.

Deufel, J., 1965. Pigmentierungsversuche mit Canthaxanthin bei Regenbogenforellen. Arch. Fischereiwiss. 16:125-132.

Dumas, R.F., 1961. Effect of light, diet, and age of spawning brown trout upon certain characteristics of their eggs and fry. New York Fish and Game J. 8:49-56.

Gutsell, J.S., 1940. The feeding of trout broodstock. Prog. Fish-Cult. 50:19-28.

Hardy, R.W., Shearer, K.D., and King, I.B., 1984. Proximate and elemental composition of developing eggs and maternal soma of pen-reared coho salmon (*Oncorhynchus kisutch*) fed production and trace element fortified diets. Aquaculture, 43:147-165.

Harris, L.E., 1984. Effects of a broodfish diet fortified with canthaxanthin on female fecundity and egg color. Aquaculture, 43:179-184.

Harris, L.E. and Griess, L.J., 1978. Trout nutrition and disease studies. Job Progress Rept. F-28-R-14. Dept. Natl. Res., Colorado Div. Wildlife. 45 pp.

Hilton, J.W., Cho, C.Y., Brown, R.G., and Slinger, S.J., 1979. The synthesis, half-life, and distribution of ascorbic acid in rainbow trout. Comp. Biochem. Physio. 63A:447-453.

Hirao, S., Yamada, J., and Kikuchi, K., 1955. Relation between chemical constituents of rainbow trout eggs and hatching rate. Bull. Jpn. Soc. Sci. Fish. 21:240-243.

Kinumaki, T., Sugii, K., Iida, H., and Takahashi, T., 1972. Addition of fat soluble vitamins to the feeding stuffs of parent rainbow trout with particular reference to the effect on the vitamin levels of eggs and fry. Bull. Tokai Reg. Fish Res. Lab. 71:133-160.

Love, R.M., 1970. The chemical biology of fishes. Academic Press, Inc., Ltd., London. 547 pp.

Morrison, J.K., and Smith, C.E., 1981. The effect of dietary carotenoid supplementation on egg quality of brook trout. Fish Development Center Infor. Leaflet No. 19. U.S.F.W.S., Bozeman, Montana. 6 pp.

National Research Council, 1981. Nutrient requirements of cold-water fishes. Washington, D.C. National Academy Press.

Orr, W., Call, J., Brooks, J., Holt, E., and Mainwaring, J., 1982. Effect of feeding rates on spawning performance of two year old rainbow trout brood stock. Fish Development Center Infor. Leaflet No. 24. U.S.F.W.S., Bozeman, Montana. 6 pp.

Phillips, A.M., Jr., Hammer, G.L., Edwards, J.F., and Hosking, H.F., 1964. Dry concentrates as complete trout foods for growth and egg production. Prog. Fish-Cult. 26:155-159.

Quantz, G., 1980. The influence of carotenoid-enriched dry feed on the egg fertilization of rainbow trout (translation of German title). Archiv. fur Fischereiwissenschaft. 31:29-40.

Reinitz, G., Orme, L., Lemm, C., and Hitzel, R., 1977. Evaluation of a shrimp supplemented diet for rainbow trout (*Salmo gairdneri*) broodstock. Final Report No. FH 1404. U.S.F.W.S., Diet Development Center, Spearfish, South Dakota.

Ridelman, J.M., 1981. Effects of starvation and diet formulation on ovarian development and egg viability of steelhead x rainbow trout hybrids. M.S. Thesis, University of Washington, 84 pp.

Ridelman, J.M., Hardy, R.W., and Brannon, E.L., 1984. The effect of short-term starvation on ovarian development and egg viability in rainbow trout (*Salmo gairdneri*). Aquaculture, 37:133-140.

Roley, D.D., 1983. The effect of diet protein level, feeding level, and the rearing water temperature on the growth and reproductive performance of rainbow trout broodstock. Ph.D. Dissertation. University of Washington, Seattle, 279 pp.

Sandnes, K., Ulgenes, Y., Braekkan, O.R., and Utre, F., 1984. The effect of ascorbic acid supplementation in broodstock feed for reproduction of rainbow trout (*Salmo gairdneri*). Aquaculture, 43:167-178.

Satia, B.P., 1973. Growth and reproductive responses of rainbow trout (*Salmo gairdneri* Richardson) to different diets. Ph.D. Dissertation. University of Washington, Seattle, 98 pp.

Scott, D.P., 1962. Effect of food quantity on fecundity of rainbow trout *Salmo gairdneri*. J. Fish. Res. Bd. Can. 19:715-730.

Smith, C.E., Osborne, M.D., Piper, R.G., and Dwyer, W.P., 1979. Effect of diet composition on performance of rainbow trout brood stock during a three-year period. Prog. Fish-Cult. 41:185-188.

Tacon, A.G.J., 1981. Speculative review of possible carotenoid function in fish. Prog. Fish-Cult. 43:205-208.

Takeuchi, T., Watanabe, T., Agino, C., Saito, M., Nishimura, K., and Nose, T., 1981. Effects of low-protein, high-calorie diets and deletion of trace elements from a fish meal diet on reproduction of rainbow trout. Bull. Jpn. Soc. Sci. Fish. 47:645-654.

Torrissen, O.J., 1984. Pigmentation of salmonids—effect of carotenoids in eggs and start-feeding diet on survival and growth rate. Aquaculture, 43:185-194.

Yu, T.C., Sinnhuber, R.O., and Hendricks, J.D., 1979. Reproduction and survival of rainbow trout (*Salmo gairdneri*) fed linolenic acid as the only source of essential fatty acids. Lipids. 14:572-575.

ABSTRACTS–Nutrition

Proximate and Elemental Composition of Developing Eggs of Pen-reared Coho Salmon *(Oncorynchus kisutch)* Fed Production and Trace Element Fortified Diets

R. W. Hardy, University of Washington and National Marine Fisheries Service; K. D. Shearer, National Marine Fisheries Service; and I. B. King, University of Washington, Seattle

Two groups of coho salmon, raised in marine net-pens, were fed standard Abernathy diet 19-2 or Abernathy diet 19-2 fortified with additional cobalt, copper, iron, manganese, and zinc during the 6-month period before spawning. Monthly ovary samples and whole fish minus the ovary samples from each group were analyzed for proximate and elemental composition. Whole body wet weights increased about 30% during the experimental period but most of the gain was in ovary weight. Analysis of the proximate changes showed a reduction in lipid levels of the fish during ovarian development. Elemental levels of the fish did not decrease. Proximate constituents were deposited in the eggs throughout ovarian development but the rates of deposition varied between constituents from month to month. Elemental deposition in the ovaries was continuous but the rates of deposition varied between elements. No evidence of maternal somatic elemental depletion was observed, and no differences were detected between the two groups in elemental composition of the maternal tissues or ovary at spawning. Analysis of the results of this study and comparisons with observations from pen-reared and wild coho salmon eggs in previous years indicated that additional elemental fortifications of Abernathy diet 19-2 are unnecessary for pen-reared coho salmon.

Mineral Supplementation of Atlantic Salmon Broodstock Diets

H. G. Ketola, U.S. Fish and Wildlife Service, Cortland, NY

Spawning of wild-kelts reconditioned on hatchery diets and of hatchery reared Atlantic salmon *(Salmo salar)* has been much less successful than that for sea-run salmon in various recent attempts in hatcheries in northeastern states. The "eye-up" of eggs from sea-run salmon ranges from 83 to 92%, in contrast to that for reconditioned salmon which ranges from 45 to 75% and that for hatchery-reared salmon, 17 to 80%. Chemical analyses of eggs and diets suggested that mineral deficiency could be the cause of poor eye-up.

A 3-month pilot study was conducted to test the effect of feeding a fish meal-containing diet with and without a mineral mixture formulated to provide sodium chloride at 0.75% of diet and magnesium, zinc, manganese, iron, copper, and iodine at 400, 150, 100, 100, 5, and 8 mg/kg of feed, respectively. Each diet was fed at about 0.5% of body weight daily to duplicate lots (40 fish each) of mixed-sex, domestically-reared 4 3/4 year-old Atlantic salmon.

The results showed that the mineral supplement had no significant ($P>0.05$) effect on egg production, fertilization rate, eye-up, hatchability or swim-up. However, there was considerable variability between replicate for all of these measurements. Semen of salmon was microscopically active and contained about 13.4 billion sperm/ml. Spermatocrit values (packed sperm cell volume) were not significantly affected by diet, nor were hematocrit values for spent females. Chemical analyses of various minerals in eyed eggs and fresh semen showed no appreciable effect of diet on any of the minerals measured.

A recent 30-month study on mineral supplementation (Zn, Mn, Fe, Cu) of fish meal diets fed to broodstock rainbow trout showed that supplementation with minerals dramatically increased eye-up and hatchability (Takeuchi et al., 1981, Bull. Jap. Soc. Sci. Fish. 47:645). This suggests that a long-term study may be essential to show any effect of supplemental minerals on reproduction in Atlantic salmon. Therefore, the lack of significant effects in this short-term study should not be considered as evidence for the non-essentiality of supplemental minerals in fish meal diets. A long-term experiment appears to be needed.

Pigmentation of Salmonids: Effect of Carotenoids in Eggs and Startfeeding Diet on Survival and Growth Rate

Ole Torrissen, Institute of Marine Research, Direktorate of Fisheries, Matredal, Norway

Salmonids mobilize their carotenoid pigments, astaxanthin and cantaxanthin, in the flesh and deposit them in eggs and skin during sexual maturation. These active metabolisms of carotenoids indicate that they have a specific function either during reproduction, early life, or both.

Carotenoids are reported to enhance growth rate, maturation rate and fecundity and to reduct mortality rate. However no biological function of astaxanthin and cantaxanthin have so far been documented by adequate scientific data.

To study the effect of carotenoids on survival, differently pigmented Atlantic salmon *(Salmo salar)* eggs and alevins were exposed to light of different wavelengths. Differently pigmented eggs and alevins incubated in darkness have also been studied. Highly pigmented eggs were found to be sensitive to light. In darkness there was no significant effect of the carotenoid level on survival of eggs and alevins. Diets supplemented with synthetical astaxanthin and cantaxanthin promoted growth rate during the early startfeeding period.

Effect of Nutritional Quality of Broodstock Diets on Reproduction of Rainbow Trout and Their Egg Quality

Takeshi Watanabe, Tokyo University of Fisheries, Tokyo, Japan

Economically productive aquaculture is heavily dependent upon an adequate supply of seed, of fertile eggs and juvenile fish, with which to stock the ponds, enclosures and other cultivation systems. One of the most important and fundamental approaches to the artificial seed production to satisfy the ever-growing demand of fish breeders is to ensure a year-round, rather than a seasonal, supply of enough fertile eggs with high qualities which produce higher survival and growth rates than those naturally occurring.

Nutrition is known to have a profound effect upon gonadal growth and fecundity. Although precise information on the nutritional requirements for gonadal maturation in broodstock is lacking, it has been generally agreed that quality and quantity of the feed, as well as the feeding regimen, are important for reproduction and egg quality. Thus this study was conducted in order to develop an adult fish diet suitable for reproduction of rainbow trout by ensuring a relationship between feed quality and egg quality.

Adult or fingerlings of rainbow trout were fed on various test diets for 3 months or 3 years to examine the effects of low protein-high calorie or EFA-deficient diets and the total deletion of trace elements from the mineral mixture in white fish meal diets on reproduction and chemical components of eggs produced.

Eggs produced from the fish fed on the low protein diet with a high energy value gave good yield of eyed eggs with high hatchability compared to those fed on the control commercial diet. But, eggs from the fish fed on the diet without supplement of trace elements or EFA were significantly low in both percentages of eyed eggs and hatchability. These results have demonstrated that the low protein-high calorie diets supplemented with beef tallow have no adverse effects on reproduction of rainbow trout, and that a supplement of trace elements to fish meal diets is indispensable for reproduction, although white fish meal contains various kinds of minerals. The same kinds of results were also obtained in red sea bream.

Effect of Ascorbic Acid Supplementation in Broodstock Feed on Reproduction of Rainbow Trout *(Salmo Gairdneri)*

Kjartan Sandnes, Yngve Ulgenes, Olaf R. Braekkan, and Finn Utne, Institute of Nutrition, Directorate of Fisheries, Nygardstangen-Bergen, Norway

Two experimental diets differing in supplementation of ascorbic acid and a third commercial diet were fed to rainbow trout broodstock. A supplementation level of 115 mg ascorbic acid per kg significantly increased the number of hatching eggs compared to eggs from fish without dietary ascorbic acid supplementation. Fish reared on the commercial diet gave eggs of similar quality as from the ascorbic acid supplemented experimental feed. The results indicate that ascorbic acid is essential for reproduction in rainbow trout. Broodstock fish should be fed adequate amounts of the vitamin to provide eggs with more than 20 μg ascorbic acid per gram.

Effects of a Broodfish Diet Fortified with Canthaxanthin on Female Fecundity and Egg Color

Larry Harris, Colorado Department of Natural Resources, Fort Collins

A synthetic form of canthaxanthin was added at 20 and 40 mg per kg of feed to a broodfish diet to determine the effect on fecundity and egg color of 3-year-old female rainbow trout *Salmo gairdneri*. The diets were fed for a 3- and 6-month period prior to spawning. Fish were checked for ripeness and spawned on a weekly basis. The egg-producing potential of the broodfish was not affected by the canthaxanthin-fortified diets. Egg color increased in direct proportion to the amount and length of time canthaxanthin was fed to the fish. The relationships between vitamins A and E with canthaxanthin may warrant study in hopes of improving female fecundity without being cost prohibitive.

The Effect of Dietary Vitamin E on the Distribution of α-Tocopherol in Rainbow Trout *(Salmo gairdneri)* During Ovarian Maturation

I. King, R. W. Hardy, and J. E. Halver, University of Washington, Seattle

Experiments were conducted to determine the effects of the presence or absence of dietary vitamin E on α-tocopherol tissue distribution in rainbow trout during egg maturation. In addition, the influence of dietary Vitamin E on ovarian development, spawning date and egg hatchability were studied. Two groups of 2+ pretreated female rainbow trout (mean initial weight 598g) were fed purified basal diets from July until spawning (January-February). Group I was fed a diet containing 90 mg dl-α-tocopheryl acetate/kg diet. Group 2 received α-tocopherol-free diet. At one or two month intervals, five fish from each group were sacrified for weight measurements and tocopherol analyses. The α-tocopherol levels in plasma, liver, muscle, and eggs were determined using high pressure liquid chromatography.

Analysis of the data showed no differences in growth, prespawning mortality rate, egg development, or egg hatchability between the two dietary treatments. α-tocopherol levels, expressed on a per gram wet basis, significantly differed between treatments in all tissues only during December. In both treatments, liver α-tocopherol levels decreased most drastically during the early months of the study. The concentration of α-tocopherol in the eggs varied at different samplings, corresponding to the absolute weight changes of the egg. α-tocopherol values in the muscle remained constant during the experimental period. Plasma α-tocopherol values reflected liver levels, but at lower concentrations. These results indicate that the fish had sufficient tissue stores of α-tocopherol at the start of the study to supply the needs of the developing ovaries without additional dietary supplementation.

Crawfish Waste—A Domestic Commercial Source of Astaxanthin

Samuel P. Meyers and Huei-Mei Chen, Louisiana State University, Baton Rouge, LA

Lousiana crawfish *(Procambarus clarkii)* heat-processed waste has been identified as a unique source of naturally-occurring biologically-active pigment based on production and availability of 30 million lbs waste/year with noteworthy pigment concentration (153 μg/g). A pilot plant has been developed for efficient pigment extraction using a vegetable or fish oil for recovery of the oil-soluble pigment. Efforts have included optimization of extraction efficiency and oil recovery, and assessment of process parameters for pending establishment of a commercial plant. Scale-up studies indicate that for every 100 lbs of crawfish waste recovered, approximately 10 lbs of astaxanthin-enriched oil (>600 ppm) can be produced. Initial projections are for a 100-250 metric ton facility. Monitoring of process parameters has allowed increase in pigment concentration to as high as 800-850 ppm. Performance trials with a variety of aquatic species (rainbow trout, coho salmon, American lobster) have demonstrated significant pigment transfer to integument and muscle.

The Effect of Diet Protein Level, Feeding Level and the Holding Water Temperature for Rainbow Trout Broodstock on Their Growth and Reproductive Performance

D. D. Roley, University of Washington, Seattle

Two separate feeding trials were conducted using University of Washington rainbow trout broodstock to determine the effect of dietary protein level, ration size and water temperature on pre-spawning growth and reproductive performance. During the eight months prior to spawning four isocaloric diets with 27, 37, 47 or 56% protein were fed to separate groups of the 1973 brood. Four groups of the 1974 brood were reared under the cross-classified design of two water temperature profiles, cool and warm; and two ration sizes, repletion and half-repletion (as a percent of body weight per day).

The dietary protein requirement for maximum growth was between 37 and 47% of a diet containing 2.8 kcal/g metabolizable energy. Caloric utilization, food conversion and protein efficiency ration decreased with increasing levels of dietary protein. Dietary protein level did not affect pre-spawning mortality, spawning success or the duration of spawning. Dietary protein level did not have a significant effect on the absolute or relative number of eggs spawned, relative egg size or embryo survival.

Maximum growth was achieved by feeding repletion rations under warm water conditions. Food and protein utlization for growth was optimum with repletion rations in warm water or half-repletion rations in cool water. Feeding repletion rations in warm water increased pre-spawning mortality. Water temperature or feeding level did not affect spawning success, but feeding half-repletion rations increased the duration of spawning and warm water delayed the time of spawning. Relative fecundity was increased by feeding half-repletion rations in warm water. Relative egg size was increased by feeding half-repletion rations in cool water. Warm water and feeding repletion rations had deleterious effects on embryo survival due to their effects on egg size; and ration size had an independent effect on embryo survival.

The First Thirty Days of Feeding Salmon and Trout

R. E. Noble, Salmon/Trout Advisory Service, Olympia, Washington

The success or failure of a given population of salmon or trout can be established in the methods of incubating fry and the first 30 days of feeding. The old practice of incubating yolk sack fry without a substrate of some type and/or using too much water flow results in a swim-up fry that must overcome a penalty. Feeding procedure and care is nearly as important as the feed. At a water temperature of 10°C salmonids should gain over 200 percent of their initial starting weight in the first 30 days of feeding. It is cost effective to assign the best fish culturist on staff to care for the fry from the day the eggs hatch to 30 days post ponding. Feed levels should be at 7 to 8% body weight and the food presented at 10-to-15 minute intervals throughout the daylight hours. A start trough having a cubic capacity of approximately 50 feet with water flows of nearly 50 gpm provides the most suitable starting environment. Size variation, disease susceptibility and feeding response along with long term survival rates are all greatly influenced in the first 30 days of feeding. Poor care results in poor fish.

The Effects of Diet on Quality of Coastal Cutthroat Trout *(Salmo clarki clarki)* Broodstock Adults, Eggs, and Fry

Tim Unterwegner, Oregon Department of Fish and Wildlife, Clackamas, Oregon

Coastal cutthroat trout *(Salmo clarki clarki)* brood stock were fed four different diets to determine the effects of diet on quality of adult brood trout and the resulting eggs and fry. The four tests diets were: Spearfish GR-3, Silvercup brood, and two formulations of Oregon Pellets. Ripe females were spawned January–March 1982 as 3-year-olds and January–March 1983 as 4-year-olds. There were significant differences ($P<0.05$) in number of eggs per female, number of eggs per mm fork length, and fry mortalities for fish spawned as 3-year-olds and differences in egg volume, number of eggs, eggs per mm fork length, and fry mortalities for fish spawned as 4-year-olds. The formal feeding trial revealed that, under the conditions of this test, the production formula of Oregon Pellets produced significantly larger fish than either Spearfish GR-3 or Silvercup, better feed conversions than Silvercup, and fewer adult mortalities than Spearfish. Standard formula Oregon Pellets contains cottonseed meal which has been shown to cause liver cancer, so we hesitate to recommend feeding it to brood stock until more research is completed.

Broodstock Management & Husbandry

Session Leader

Lauren R. Donaldson

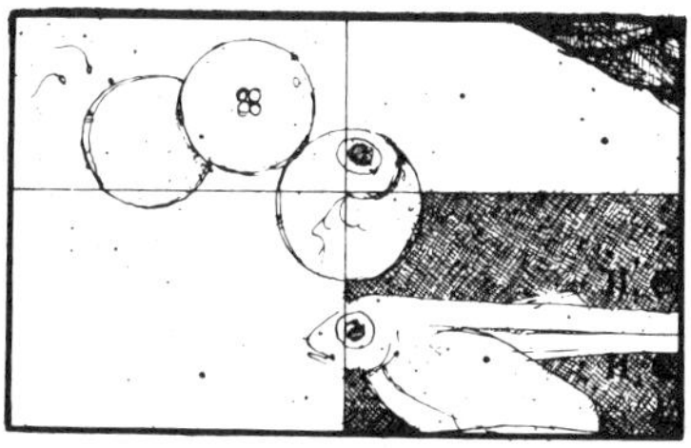

ARTIFICIAL INSEMINATION IN SALMONIDS

Rolland Billard[1]

Artificial insemination is generally used in salmonids whether the fish are raised for food or for ranching. Even after a century of practice, artificial insemination is still carried out in a traditional way, i.e. ovulation in females is checked once or twice a week, necessitating frequent handling of non-ovulated fish, and the eggs of several females are mixed with sperm taken directly from one or two males. One male is usually used to fertilize several females. Sometimes the fish farmer takes males from among the fish that have reached commercial size (150-250 g), but this procedure favors early puberty. Generally, the fish farmer has a male brood stock in reserve which includes 20% to 40% of the brood biomass. The number of males in a fish farm is thus much more critical than it is in a farm where domestic animals are raised. Although there is a market for large-sized fish, culled males have little commercial value because their quality is altered (Bye and Lincoln, 1979).

The present paper, a review of published and unpublished works from our laboratory and others shows that artificial insemination techniques can be simplified and the number of fish-farm males reduced by better utilization of their sperm when diluted in an appropriate diluent. More detailed information on artificial insemination and the management of brood-fish gametes is available in reviews by Billard (1978, 1980a), Scott (1981) and Scott and Baynes (1980).

Materials and Methods

The standard procedures used in the laboratory and employed in the various experiments reported here have already been described by Billard et al. (1974a,b). Briefly, sperm fertilizing ability and ovum fertilizability are tested by exposing the sperm or ova to various temperatures and media over a period of time, after which they are mixed with fresh sperm or ova in an artificial insemination diluent (D532) (Billard, 1977a, see below). Batches of about 200 eggs each are usually diluted with 10 ml of diluent, and sperm is added at a dilution rate of 10^{-2} to 10^{-5} (100 to 0.1 μl). The eggs are then incubated at 10°C for 100 or 200 days (eyed-stage) or until hatching; the percentage of developed eggs is the criterion of successful fertilization. In some cases, the abnormal embryos are counted. Statistical comparison is made using the t-test or variance analysis after angular transformation of the percentages. The values are plotted ± SD or with a 95% confidence interval (percent) ± CI (Snedecor and Cochran, 1966).

Spermatogenetic Production and Spermiation Yield

In salmonids, the concentration of spermatozoa in the testis is very high: 58×10^9 spermatozoa/g testis in rainbow trout (Billard et al., 1971); 55×10^9 in Atlantic salmon (Kazakov, 1979); and 25×10^9 in brown trout (Billard, 1983a). Since testis weight is about 8-13% of total body weight in adult salmonids, the

[1] Laboratoire de Physiologie des Poissons, I.N.R.A., France

total spermatogenetic production, i.e. the total number of spermatozoa produced during the reproductive cycle, reached 5-7x10^9/g body weight/year. Recorded values are less in young fish: about 1x10^9 spermatozoa/kg body weight/year in rainbow trout (Billard, 1974) and brown trout (Billard, 1983a; in this paper read 1x10^9 instead of 6.8x10^7). Spermatogenetic yield varies greatly among species; some fish produce more than 10 times more spermatozoa than others, and the differences are even greater if sperm yield is related with female fecundity. The number of spermatozoa per egg laid by the female is 1.6x10^9 in *Leporinus* (a South American species), 20x10^9 in pike and 3500x10^9 in rainbow trout (Billard, 1982a,b). This may be due to the mode of reproduction; the overall amount of sperm required is lower then the fish spawn in pairs (pike) than when they spawn in groups (salmonids) (Billard, 1982c). In trout, the number of spermatozoa available for artificial insemination is much less than the number produced. First, the period of spermiation far exceeds the period of ovulation: the males begin to produce sperm before the females start to ovulate and sperm production continues after ovulation has ended. Secondly, all the spermatozoa produced are not released during the period of spermiation; they remain in the testis and are resorbed (Billard et al., 1971). The spermiation yield of fish in captivity has been extimated at 22% and 40% in rainbow trout (Billard et al., 1971; Billard, 1974) and at 15% in brown trout (Billard, 1983a).

In conclusion, spermatogenetic production is high in salmonids but spermiation yield, i.e. the volume of milt available for artificial insemination, is poor. The reason for this may be a failure in the endocrine control of spermiation (Scott et al., 1980; Billard et al., 1982; Fostier et al., 1982, 1983). Improvement may be obtained by placing males in optimal environmental conditions, including a low temperature (Kime, 1979; Billard and Breton, 1977) and the presence of females (Colombo et al., 1982; Liley, 1982). If needed, hormonal treatment can be used to stimulate spermiation (Billard et al., 1982; Weil and Crim, 1982, 1983; Crim et al., 1983). In the actual practice of salmonid reproduction, 1000 to 5000 billion spermatozoa are produced by a 1-kg male and about 205 (200-1000) billion are available for artificial insemination.

Gamete Survival

Criteria for estimating gamete quality: There are no true, dependable criteria for estimating gamete quality. In salmonids, the length and intensity of spermatozoon motility are not invariably correlated with fertilizing ability (Goryczko and Tomasik, 1975), especially as concerns sperm stored *in vitro* for several days or deep-frozen sperm (Legendre and Billard, 1980). This may be due to the difficulty in objectively assessing sperm motility. The elastic light scattering technique used for guppy (Berge et al., 1967) and trout (Moccia and Munkittrick, 1984; Billard and Breton, 1976) is promising because it seems more closley related to fertilizing ability. Also, the Na^+/K^+ ratio in the seminal fluid, proposed by Hwang and Idler (1969), does not change in parallel with fertilizing ability when the sperm is stored for several days (Carpentier and Billard, 1978).

Other parameters, such as pH, osmotic pressure and ionic composition (Nomura, 1964) may be implicated but have not been correlated with fertilizing ability. In trout farming, the motility criterion can only be employed in a negative way: fresh sperm of high fertilizing ability cannot be identified and therefore selected, while that of weak motility, which may have a low fertilizing ability, is eliminated at insemination. The method of Winnicki and Tomasik (1976) for assessing sperm quality consists in measuring the swelling of spermatozoa after they have been diluted in media of increasing salinity.

Morphological criteria are used to determine ovum quality. Layings having a high percentage of considerably hydrated ova of heterogeneous aspect with marked polarization of the lipid droplets or the yolk should be rejected because they will give a low fertility rate. The quality of trout ova has been assessed according to whether the ova are mature or over-mature (aged) (Nomura et al., 1974).

***In vivo* survival**: Spermatogenesis is very seasonal in salmonids. The spermatozoa are all formed at nearly the same time and are released over a period of several months during which they age. On a short-term basis, sperm stored in the spermiduct or in the testis does not generally vary much over a 14-day period, although there may be some individual variation (Billard et al., 1981). Preliminary observations in rainbow trout have shown that the quality of the sperm, i.e. its potential for dilution or cryopreservation or fertilizing ability, decreases as the spermiation period advances (Billard, 1979 and unpublished data). As increasing the number of spermatozoa used for insemination may compensate for decreased quality (Carpentier and Billard, 1978), the volume of milt used for artificial insemination must be higher at the end of spawning than at the beginning (see below). Similar results were obtained by Matei et al. (1980). In contrast, changes are more pronounced at the end of the spawning period; Billard (1976a) has shown that the first ml of milt taken is of poor quality and must be discarded. Temperature (Billard and Breton, 1977) and salinity (Sower, personal communication, Stoss and Fagerlund, 1982) also influence sperm quality.

In the female, ova remaining in the body cavity after ovulation become overmature; good survival of 30 days after ovulation has been reported in large rainbow trout females with more than two reproductive cycles (Escaffre et al., 1976, 1977), while in younger females survival is shorter (8 to 15 days) (Escaffre et al., 1977; Escaffre and Billard, 1979). Sakai et al. (1975) observed post-ovulatory ovum survival of 1 to 2 weeks in rainbow trout. Interfemale variability of egg survival in brown trout increased over a 72-hour period (Billard et al., 1981); when females died accidentally in water, their eggs survived only a few hours (Billard et al., 1981; Matei et al., 1980).

The factors causing variability in ovum survival after ovulation are not known. It has been shown that plasma GTH is higher in rainbow trout females retaining ova than in spent individuals (Jalabert and Breton, 1980), but this high GTH level is not correlated with ovum survival (Figure 1). The survival of ova may depend on internal factors, such as number of reproductive cycles, physical conditions, quantity and quality of food during gametogenesis, and on external factors such as temperature (Billard and Breton, 1977) or salinity (Stoss and Fagerlund, 1982). As the effects of internal factors as well as egg quality are difficult to predict, and females should be checked for ovulation rather frequently to avoid overmaturation, i.e. around once a week in rainbow trout kept at a maximum termprature of 10°C.

Gamete survival after sampling and possibilities of short-term preservation when undiluted: Immediately after collection, salmonid sperm can be stored un-

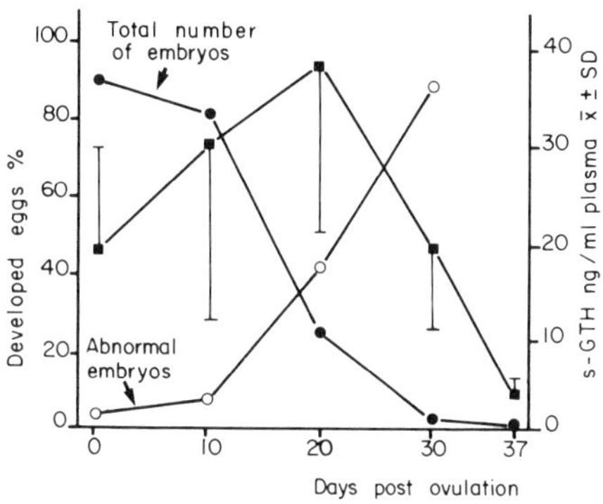

Figure 1. Post-ovulatory changes in plasma GTH and ovum survival in rainbow trout measured by the percent of normal and abnormal developing eggs at hatching. (From Marcel et al., unpublished.)

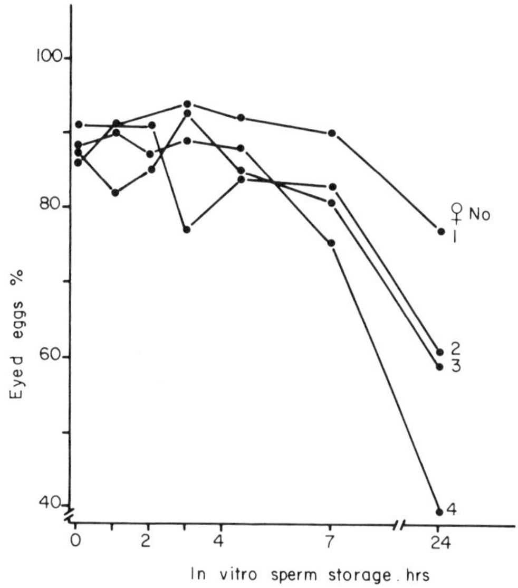

Figure 2. Changes in the fertilizing capacity of trout sperm stored at 10°C for 24 hours. Aliquots of pooled sperm were taken at various intervals to fertilize the eggs of four females. Dilution rate: 10-3.

diluted from several hours to several days at a temperature above 0°C (Tomasik, 1974; Carpentier and Billard, 1978). In the experiment shown on Figure 2, the fertilizing ability of sperm stored in a large vial remained at its initial level for 4 hours; after 24 hours, this ability decreased significantly. Longer survival may be obtained by adding some cryoprotectants and storing the sperm at sub-zero temperature (Truscott et al., 1968; Sanchez-Rodriguez and Billard, 1977) or by adding oxygen or antibiotics (Billard, 1981; Stoss and Holtz, 1983; Stoss and Refstie, 1983). Survival is usually better when the sperm is stored under a thin milt layer of several mm which supposedly oxygenates the milt better.

After collection, the ova can be stored at temperatures of 0.5°C or more, but they have never been frozen successfully. The numerous studies carried out have shown very different storage times, depending on the individuals or the experiment. Ova are not usually diluted for storage but simply left in the coelomic fluid. Survival is short at room temperature: between 2 and 8 hours at 18°C, depending on the female (Billard and Breton, 1977), and about 24 hours at 10°C (Withler and Morley, 1968). The ova may be stored in a refrigerator for several days (Barrett, 1951; Takano et al., 1973), and at 0°C, their fertilizability remains high for 7 days in brown trout and for 3 days in rainbow trout (Carpentier and Billard, 1978). Ova have also been stored in sacks inflated with 0_2 (Hiroi, 1978; Billard, 1981).

In conclusion, these results show that sperm can be stored (with possible addition of oxygen) for several hours after collection. This allows it first to be collected and then to be used as needed on the female ova.

Survival after dilution: At the time of artificial insemination, the gametes are usually mixed together and diluted with water. The role of such a diluent is to initiate the motility of all the spermatozoa and to keep the ratio of ova to spermatozoa optimal so that all fertilizable ova will be fertilized with a minimal number of spermatozoa. It has therefore been necessary to develop a diluent suitable for both gametes which would allow the highest rate of fertilization.

A diluent for rainbow trout proposed by Nomura in 1964 was based on diluted sea water. Diluent parameters were defined more precisely by Petit et al. (1973), Billard et al. (1974a,b), Billard and Jalabert (1974) and Billard (1975a,b; 1977a,b). Finally, it was found that a simple NaCl solution, buffered at pH 9 with Tris 20 mM and glycine 50 mM, final osmotic pressure 250 mOsm was very favorable to egg and sperm survival and gave a consistently higher percentage of fertilization than water. In this saline solution, called D532, the swelling and rupture of the spermatozoon plasma membrane is prevented (Billard, 1978, 1983b) and motility is slightly prolonged due to the fact that the micropyle is not sealed as it is soon after egg contact with fresh water (Szöllösi and Billard, 1974). The saline solution also prevents the precipitation of egg yolk resulting from eggs accidentally crushed during stripping. In water, the precipitated yolk traps the spermatozoa, hampering their movement and perhaps also plugging the micropyles.

D532 is as beneficial to ovum survival after dilution as other media such as a diluent mimicking the mineral composition of coelomic fluid or seminal plasma (Billard and Jalabert, 1974). The problem as to whether coelomic fluid is a good candidate for artificial insemination diluent has been widely discussed in the past; coelomic fluid volume and quality are not constant, and its composition may vary from one female to another and during egg retention in the post-ovulatory period (Satia et al., 1974; Czihak et al., 1979; Cetta and Goetz, 1982). Coelomic fluid may also be diluted with water (Fish and Ginnelly, 1966). Some females may show signs of hydropisia and their cavity fluid is not suitable for fertilization (Dorier, 1949). As coelomic fluid does not appear to be better than properly buffered mineral diluent when used in artificial insemination with intact gametes (Billard, 1983c), it may be replaced easily by D532 (defined above) which is simple and readily available when performing artificial insemination. Such a mineral diluent is now used in some fish farms in France. It contains a mixture of NaCl + Tris + glycine which is dissolved in hatchery water to include other ions important in fertilizaton (Gilkey, 1981) and sperm motility such as Ca^{++}.

In this diluent, sperm survival is only slightly improved (Figure 2) and we have tried to prolong sperm fertilizing ability. Adding BSA into the diluent at a dose of 10 mg/ml slightly improves survival, but higher doses do not improve it further (Figure 3). When cAMP (dibutyril form) is added to the diluent, survival is markedly prolonged (Figure 3). This is in agreement with the finding of Benau and Terner (1980) and indicates that the reduction of motility in rainbow trout sperm during the spawning season is correlated with a decline in cAMP concentration in the spermatozoa. Prolongation of motility has been also obtained with theophylline (Billard, 1980b). It is therefore possible to produce a more complex diluent for artificial insemination which would be useful in case sperm quality is deficient: a) after short-term storage or at the end of the spawning season; b)

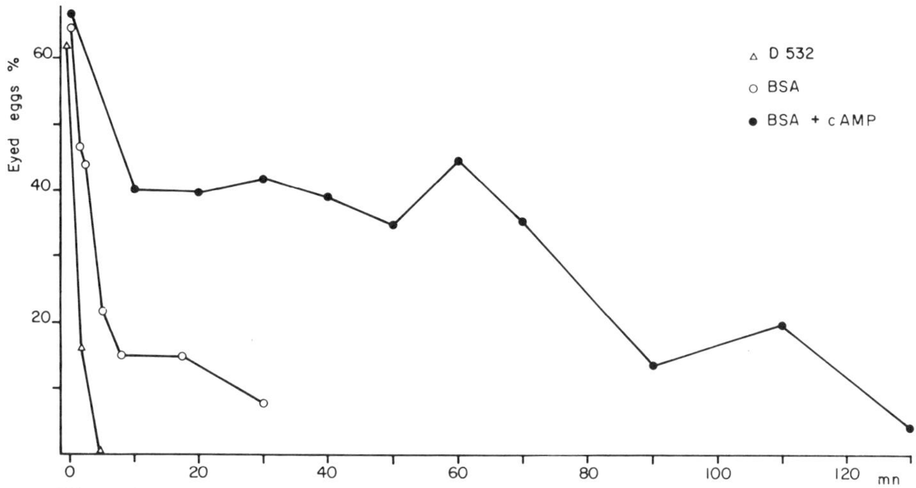

Figure 3. Sperm survival after dilution in diluent along or with 10 mg/ml BSA or BSA 10 mM cAMP (dibutyril) (rainbow trout). Dilution rate: 10-3.

sperm originating from males kept in unfavorable conditions as high temperature or salinity; c) cryopreserved sperm which is very fragile after thawing. Another way of improving gamete quality is to offer the brood stock better environmental conditions, including a better diet; this allowed Baynes and Scott (1982) to cryopreserve rainbow trout sperm more efficiently.

Definition of the Optimal Egg-Spermatozoa-Diluent Ratio for Artificial Insemination

Only a few studies have tried to define the minimum number of spermatozoa needed to fertilize one egg. This number varies between 10,000 and 300,000 (Billard et al., 1974a; Billard and Carpentier, 1974; Billard, 1975b). Such values look high, considering that the gametes are put in direct contact with each other. It should be pointed out that survival is short and the distance the spermatozoon swims during its lifespan is limited (about 2 mm according to Schlenk and Kahmann, 1938), being much less than egg diameter (~16 mm in trout). As the spermatozoon can enter the egg only via the micropyle, a single spermatozoon has little chance of reaching its target, and it is necessary to use a large number of them. Wide variation in the egg-spermatozoon ratio is likely to result in changes in sperm quality, especially in motility, intensity and duration and in the distance covered by the spermatozoon. Sperm of poor quality is also very sensitive to high dilution: at 10^{-4} fertilization success is variable but it is consistently better at 10^{-3}.

Ovum dilution should also be considered. Billard (1975b) showed that there is an optimal spermatozoon-egg-diluent ratio. At an egg concentration of 10-20/ml of diluent, fertilization rate is high with a minimal number of spermatozoa.

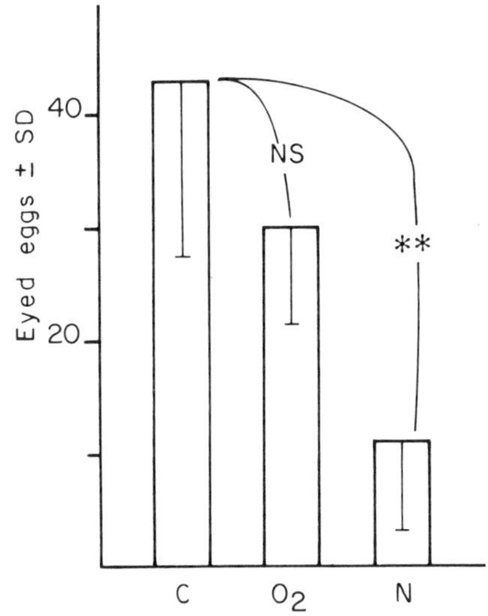

Figure 4. Changes in egg fertility after insemination in normal conditions (C) or with bubbling oxygen (O2) or nitrogen (N). Values given are X ± SD. n = 3. NS: non significant, ** $P < 0.01$.

Definition of Some Optimal Conditions for Artificial Insemination

Some important factors in salmonid insemination are still to be defined. The effect of temperature has been studied by Hokanson et al. (1974), Billard and Gillet (1975) and Billard and Breton (1977). The optimal temperature for artificial insemination depends on the rate of sperm dilution and gamete quality. At 10^{-4} or 10^{-3} dilution, fertilization is usually more successful at 5-15°C than at 0.5° or 20°C. The oxygen requirement of the gametes at insemination has not been defined precisely. Czihak et al. (1979) observed that oxygen demand was not very high at the time of fertilization in rainbow trout. A preliminary experiment (Figure 4) comparing the results of normal insemination with insemination in D532 with bubbling oxygen or nitrogen showed that the percentage of fertilization was not improved with O^2 and was significantly decreased with nitrogen.

Gametes for artificial insemination are usually mixed with a feather; a simpler way when a diluent is used is just to pour the eggs diluent back and forth as soon as the sperm has been added.

Another question is how long after insemination the eggs should stay in the D532 + spermatozoa before they are transferred to fresh water in the incubator. It has been shown that when the eggs are poured into incubators 5 to 10 minutes after insemination, the fertilization rate is lower, probably due to some distance in the fertilization process (Billard et al., 1974a). After 10 minutes, this drop in fertility is not observed so it is advised to let the eggs stand in the diluent for 15 minutes before transferring them. In order to avoid contaminating the incubation system with dead spermatozoa, which furnish substrate for fungal development, it is better to rinse the eggs very quickly with water before transferring them. After insemination and during rinsing and hardening, the eggs are rather sensitive and should be handled carefully.

Egg sensitivity to mechanical shock has been reported (Jensen and Alderdice, 1983) and shows individual variation (Billard, 1976b); eggs are also sensitive to low temperatures around 0.5°C. Moreover, it is well to avoid exposing the gametes to direct sunlight.

Artificial Insemination Procedure

The insemination procedure using a diluent is summarized in Figure 5. As sperm can survive undiluted for some hours after collection, all males can first be sampled in the morning and the sperm pooled; this has the advantage of more extensive genetic mixing. The sperm is stored until use in a cold place (ice, refrigerator) and may be supplied with oxygen. The females are then stripped and their eggs collected in a large dishpan. The fish should be stripped separately so that poor-quality eggs which are swollen or have an heterogeneous yolk can be discarded. The eggs of several females can be pooled afterwards in a dishpan of known volume (5 or 10 liters, for instance). The excess coelomic fluid is discarded and replaced by diluent. The egg-diluent ratio is about 3 liters of eggs to one liter of diluent; just covering the eggs with the diluent is sufficient. One to 3 ml of sperm is added on top of this, and the whole mixture is immediately poured back and forth using a second dishpan. The mixture is allowed to stand for 15 minutes before quick rinsing and transfer into incubators.

Conclusion

Insemination has been practiced in salmonids since the Middle Ages. Jacobi (1758) mentions this technique and Gehin and Rémi re-discovered it in 1842 (cited by Pizzetta, 1875). In the widely used "dry" method, the eggs of several females are mixed with the sperm of several males and water is then added, activating spermatozoon motility. In a variation of this method, the coelomic fluid is retained and spermatozoon motility is activated by contact with the ova. These techniques do not give a good yield because there is an excess (some ten million) of spermatozoa. Several authors suggested reducing the amount of sperm and showed that one male can fertilize the ova of more than 20 females (Nursall and Hasler, 1952) or even of some hundred females (Morley and Withler, 1969).

Using the present technique of properly buffered diluent, a large number of eggs can be fertilized in a very simple way that is practical in trout hatcheries. To summarize, a 1-kg male yields at least 200 billion spermatozoa; with a maximum of 200,000 spermatozoa per egg when inseminating, such a male can fertil-

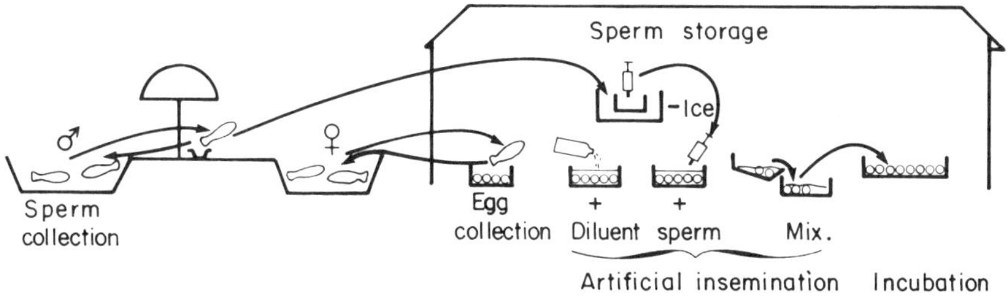

Figure 5. Diagram showing the proposed technique of artificial insemination using a diluent.

ize 1 million eggs, i.e. the eggs produced by 500 1-kg females. Such a ratio is applicable only in large hatcheries because a minimum number of males must be used to ensure sufficient genetic mixing. This technique is widely used in Japan (Nomura, personal communication).

Finally, the technique of artificial insemination with a diluent offers the following advantages:

- Better use of gametes, especially of spermatozoa, allowing a reduction in the number of hatchery males as well as increased possibility of genetic selection
- Higher fertilization rate due to slight improvement of the duration of sperm motility and to the elimination of precipitated yolk from ova crushed during sampling
- Simplified procedure, including separate handling of males and females, simple estimation of sperm-diluent ratio and easier mixing of gametes

References

Barrett, I., 1951. Fertility of salmonids eggs and sperm after storage. J. Fish. Res. Board Can. 8:125-133.

Baynes, S.M. and Scott, A.P., 1982. Cryopreservation of rainbow trout spermatozoa: variation in membrane composition may influence spermatozoan survival. In: C.J.J. Richter and H.J.Th. Goos (Comp.), Reproductive Physiology of Fish, PUDOC Wageningen, p. 128.

Benau, D. and Terner, C., 1980. Initiation, prolongation and reactivation of salmonid spermatozoa. Gamete Res., 3:247-257.

Berge, P., Volochine, B., Billard, R., and Hamelin, A., 1967. Mise en evidence du mouvement propre de microorganismes vivants grace a l'etude de la diffusion inelastique de la lumiere. C. R. Acad. Sci. Paris 265:889-892.

Billard, R., 1974. La production spermatogenetique de la truite arc-en-ciel au cours du premier cycle reproduction. Bull. Fr. Pisci. 253:139-149.

Billard, R., 1975a. L'insemination artificielle de la truite *Salmo gairdneri* Richardson. IV-Effets des ions Ket Nasur la conservation du pouvoir fecondant des gametes. Bull. Fr. Pisci. 256:88-100.

Billard, R., 1975b. L'insemination artificielle de la truite *Salmo gairdneri* Richardson. V-Effets de la dilutionet definition du rapport optimum gametes/dilueur. Bull. Fr. Pisci. 257:121-135.

Billard, R., 1976a. Variation de la qualite des gametes a differents niveaux du tractus genital chez la truite arc-en-ciel. Ann. Hydrobiol. 7:97-104.

Billard, R., 1976b. Effet de l'agitation des ovules et des oeufs de truite arc-en-ciel sur la fecondabilite et la debut du developpement embryonnaire. Bull. Fr. Pisci. 262:5-11.

Billard, R., 1977a. A new technique of artificial insemination for salmonids using a sperm diluent. Fisheries 1:24-25.

Billard, R., 1977b. Utilisation d'un systeme Tris-glycocolle le pour tamponner le dilueur d'insemination de truite. Bull. Fr. Pisci. 264:102-112.

Billard, R., 1978. Some data on gametes preservation and artificial insemination in teleost fish. Actes de colloques du CNEXO 8:59-73.

Billard, R., 1979. La gametogenese, le cycle sexuel et le controle de la reproduction chez les poissons teleosteens. Bull. Fr. Pisci., 273:117-136.

Billard, R., 1980a. Reproduction and artificial insemination in teleost fish. 9th Inter. Congress anim. Reprod. Artif. Insem. Madrid, II:327-337.

Billard, R., 1980b. Prolongation de la duree de motillite et du pouvoir fecondant des spermatozoides de truite arc-en-ciel par addition de theophylline au milieu de dilution. C. R. Acad. Sci. Paris 291:649-652.

Billard, R., 1981. Short-term preservation of sperm under oxygen atmosphere in rainbow trout (*Salmo gairdneri*). Aquaculture 23:287-293.

Billard, R., 1982a. On some patterns of reproductive physiology in male teleost fish. In: C.J.J. Richter and H.J.Th. Goos (Comp.), Reproductive Physiology of Fish, PUDOC, Wageningen, p. 192.

Billard, R., 1982b. Some aspects of spermatogenesis and spermatology in teleost fish. IVth Inter. Symp. on Spermatology, June-July 1982, Seillac.

Billard, R., 1982c. Different patterns of spermatogenesis and spermatology in teleost fish. Symp. Tactic and Strategy in Fish, July 1982, Plymouth.

Billard, R., 1983a. A quantitative analysis of spermatogenesis in the trout *Salmo trutta fario*. Cell. Tissue Res. 230:495-502.

Billard, R., 1983b. Ultrastructure of trout spermatozoa: changes after dilution and deep-freezing. Cell. Tissue Res. 228:205-218.

Billard, R., 1983c. Effects of coelomic and seminal fluids and various saline diluents on the fertilizing ability of spermatozoa in the rainbow trout *Salmo gairdneri*. J. Reprod. Fert., 68:77-84.

Billard, R. and Breton, B., 1976. Sur quelques problemes de physiologie du sperme chez les poissons teleosteens. Rev. Trav. Inst. Peches Marit., 40:501-503.

Billard, R. and Breton, B., 1977. Sensibilite a la temperature des differentes etapes de la reproduction chez la truite arc-en-ciel. Cahiers Lab. Montereau 5:5-24.

Billard, R. and Carpentier, M., 1974. Determination du nombre optimum de spermatozoides necessaires a la fecondation d'un ovule au cours de l'insemination artificielle de la truite. Bull. Fr. Pisci. 251:73-76.

Billard, R. and Gillet, C., 1975. Effets de la temperature sur la fecondation et la survie des gametes de la truite arc-en-ciel *Salmo gairdneri*. Bull. Fr. Pisci. 259:53-65.

Billard, R. and Gillet, C., 1981. Vieillissement des ovules et potentialisation par la temperature des effets des micropolluants du milieu aqueux sur les gametes chez la truite. Cahiers Lab. Montereau 12:35-42.

Billard, R. and Jalabert, B., 1974. L'insemination artificielle de la truite *Salmo gairdneri* Richardson. II-Comparaison des effets de differents dilueurs sur la conservation de la fertillite des gametes avant et apres insemination. Ann. Biol. animal. Bioch. Biophys. 14:601-610.

Billard, R., Breton, B., and Jalabert, B., 1971. La production spermatogenetique chez la truite. Ann. Biol. anim. Bioch. Biophys. 11:190-212.

Billard, R., Petit, J., Jalabert, B., and Szollosi, D., 1974a. Artificial insemination in trout using a sperm diluant. In: J.H.S. Blaxter (Editor), Early life history of fish, Springer Verlag, Berlin, Heidelberg, New York. pp. 715-723.

Billard, R., Jalabert, B., and Breton, B., 1974b. L'insemination artificielle de la truite *Salmo gairdneri* Richardson. III-Definition de la nature et de la molarite du tampon a employer avec les dilueurs d'insemination et de conservation. Ann. Biol. anim. Bioch. Biophys. 14:611-621.

Billard, R., Marcel, J., and Matei, D., 1981. Survie *post mortem* des gametes de truite fario, *Salmo trutta* fario. Can. J. Zool. 59:29-33.

Billard, R., Fostier, A., Weil, C., and Breton, B., 1982. The endocrine control of spermatogenesis in teleost fish. Can. J. Fish Aquat. Sci. 39:65-79.

Bye, V. and Lincoln, R., 1979. Male trout. An expensive luxury. Fish Farmer 2:19-20.

Carpentier, P. and Billard, R., 1978. Conservation a court terme des gametes de salmonides a des temperatures voisines de 0°C. Ann. Biol. anim. Bioch. Biophys. 18:1083-1088.

Cetta, F. and Goetz, F.W.M., 1982. Ovarian and plasma prostaglandin E and F levels in brook trout (*Salvelinus fontinalis*) during pituitary-induced ovulation. Biol. Reprod. 27:1216-1221.

Colombo, L., Belvedere, P.C., Marconato, A., and Bentivegna, F., 1982. Pheromones in teleost fish. In: C.J.J. Richter and H.J.Th. Goos (Comp.), Reproductive Physiology of Fish, PUDOC Wageningen, pp. 84-94.

Crim, L.W., Evans, D.M., and Vickery, B.H., 1983. Manipulation of the seasonal reproductive cycle of the landlocked Atlantic salmon (*Salmo salar*) by LHRH analogues administered at various stages of gonadal development. Can. J. Fish. Aquat. Sci. 40:61-67.

Czihad, G., Peter, R., Puschendorf, B., and Grunicke, H., 1979. Some data on the basic metabolism of trout eggs. J. Fish Biol. 15:185-193.

Dorier, A., 1949. Action du liquide coelomique sur les spermatozoides de truite arc-en-ciel. Trav. Lab. Hydrobiol. Pisci., Univ. Grenoble, 41:69-73.

Escaffre, A.M. and Billard, R., 1979. Evolution de la fecondabilite des ovules de truite arc-en-ciel *Salmo gairdneri* laisses dans la cavite abdominale au cours de la periode post-ovulatoire. Bull. Fr. Pisci. 272:57-70.

Escaffre, A.M., Petit, J., and Billard, R., 1976. Evolution de la fecondabilite des ovules de truite arc-en-ciel laisses dans la cavite coelomique apres ovulation. Rev. Trav. Inst. Peches marit. 40:561-562.

Escaffre, A.M., Petit, J., and Billard, R., 1977. Evolution de la quantite d'ovules recoltes et conservation de leur aptitude a etre fecondes au cours de la periode post-ovulatoire chez la truite arc-en-ciel. Bull. Fr. Pisci. 265:134-142.

Fish, G.R. and Ginnely G.D., 1966. An adverse effect of coelomic fluid on unspawned ova in trout. Trans. Am. Fish. Soc. 94:104-107.

Fostier, A., Billard, R., Breton, B., Legendre, M., and Marlot, S., 1982. Plasma 11-oxo-testosterone and gonadotropin during the beginning of spermiation in rainbow trout (*Salmo gairdneri* R.). Gen. Comp. Endocrinol. 46:428-434.

Fostier, A., Billard, R., and Breton, B., 1983. Plasma 11-oxo-testosterone and gonadotropin in relation to the arrest of spermiation in rainbow trout (*Salmo gairdneri*). Gen. Comp. Endocrinol. (submitted for publication).

Gilkey, J.C., 1981. Mechanisms of fertilization in fishes. Amer. Zool. 21:359-375.

Goryczko, K. and Tomasik, L., 1975. An influence of males on the viability and fertilization degree on trout eggs. Act. Ichthyol. Pisci., Vol. V, 1-11.

Hiroi, O., 1978. Studies on the retention of gametes of salmonid fishes. 3. Change in fry liberation rate of stored chum salmon eggs inseminated with sperms preserved in same conditions. Scient. Rep. Hokkaido Salm. Hatch. 32:19-26.

Hokanson, K.E.F., McCormick, J.H., Jones, B.R., and Tucker, J.H., 1973. Thermal requirements of maturation, spawning, and embryo survival of the brook trout, *Salvelinus fontinalis*. J. Fish. Res. Board Can. 30:975-984.

Hwang, P.C. and Idler, D.R., 1969. A study of major cations, osmotic pressure and pH in seminal components of Atlantic salmon. J. Fish. Res. Board. Can. 26:413-419.

Jalabert, B. and Breton, B., 1980. Evolution de la gonadotropine plasmatique t-GTH apres l'ovulation chez la truite arc-en-ciel (*Salmo gairdneri* R.) et influence de la retention des ovules. C. R. Acad. Sci. Paris, serie D, 290:799-801.

Jensen, J.O.T. and Alderdice, D.F., 1983. Changes in mechanical shock sensitivity of coho salmon (*Oncorhynchus kisutch*) eggs during incubation. Aquaculture 32:303-312.

Kazakov, R.V., 1979. Storing of sexual products of Atlantic salmon at low temperature in quality of parents and sexual products in fish. Gosniord, Leningrad, pp. 3-19.

Kime, D.E., 1979. The effect of temperature on the steroidogenic enzymes of the rainbow trout. Gen. Comp. Endocrinol. 39:290-296.

Legendre, M. and Billard, R., 1980. Cyropreservation of rainbow trout sperm by deep-freezing. Nutr. Repro. Develop. 10:1859-1868.

Liley, N.R., 1982. Chemical communication in fish. Can. J. Fish. Aquat. Sci. 39:22-35.

Matei, D., Marcel, J., and Billard, R., 1980. Survie des gametes en conditions *in vitro* et *post mortem* chez la truite arc-en-ciel. Bull. Acad. Sci. Agric. For. Roum. 10:211-217.

Moccia, R.D. and Munkittrick, K.R., 1984. The relationship between fertility of rainbow trout eggs and motility of spermatozoa as evaluated by quasielastic light scattering techniques. (This symposium).

Morley, R.B. and Withler, F.C., 1969. Observations on sockeye and pink salmon ova and milt relative to spawn collection. Fish. Res. Board Can. 111:1-11.

Nomura, M., 1964. Studies on reproduction of rainbow trout with special reference to egg taking. VI. The activities of spermatozoa in different diluents and perservation of semen. Bull. Jap. Soc. Sci. Fish 30:732-733.

Nomura, M., Sakai, K., and Takashima, F., 1974. The over-ripening phenomenon of rainbow trout. I. Temporal morphological changes of eggs retained in the body cavity after ovulation. Bull. Jap. Soc. Sci. Fish. 40:977-984.

Nursall, J.R. and Hasler, A.D., 1952. A note on experiments designed to test the viability of gametes and the fertilization of eggs by minute quantities of sperm. Progres. Fish. Cult. 14:165-168.

Petit, J., Jalabert, B., Chevassus, B., and Billard, R., 1973. L'insemination artificielle de la truite. I. Effets du taux de dilution, du pH et de la pression osmotique du dilueur sur la fecondation. Ann. Hydrobiol. 4:201-210.

Pizzetta, J., 1875. La pisciculture fluviale et maritime en France. J. Rothschild, Ed., 468 p.

Sakai, K., Nomura, M., and Takashima, F., Oto, H., 1975. The over-ripening phenomenon of rainbow trout. II. Changes in the percentage of eyed eggs, hatching rate and incidence of abnormal alevins during the process of over-ripening. Bull. Jap. Soc. Sci. Fish. 41:855-860.

Sanchez-Rodriguez, M. and Billard, R., 1977. Conservation de la motilite et du pouvoir fecondant du sperme de la truite arc-en-ciel maintenu a des temperatures voisines de 0°C. Bull. Fr. Pisci. 265:144-152.

Satia, B.P., Donaldson, L.R., Smith, L.S., and Nightingale, J.N., 1974. Composition of ovarian fluid and eggs of the University of Washington strain of rainbow trout (*Salmo gairdneri*). J. Fish Res. Board Can. 31:1796-1799.

Schlenk, W. and Kahmann, H., 1938. Die chemische zusamennetzung des spermaliquors und ihre physiologische bedeutung. Untersuchung am Forellensperma. Biochem. Z. 295:283-301.

Scott, A.P., 1981. Bibliography of biology, handling and storage of fish spermatozoa. Biblphy. Reprod., 37, 6 pp.

Scott, A.P. and Baynes, S.M., 1980. A review of the biology, handling and storage of salmonid spermatozoa. J. Fish Biol. 17:707-739.

Scott, S.P., Bye, V.J., Baynes, S.M., and Springate, J.R.C., 1980. Seasonal variations in plasma concentrations of 11-ketotestosterone and testosterone in male rainbow trout, *Salmo gairdnerii* Richardson. J. Fish Biol. 17:495-505.

Snedecor, G.W. and Cochran, W.G., 1966. Statistical methods applied to experiments in agriculture and biology. The Iowa State University Press, Ames, Iowa, USA.

Stoss, J. and Fagerlund, U.H.M., 1982. Influence of saltwater spawning and stress on quality of sex products on chum salmon (*Oncorhynchus keta*). In: C.J.J. Richter and H.J.Th. Goos (Comp.), Reproductive Physiology in Fish, PUDOC, Wageningen. pp. 244.

Stoss, J. and Holtz, W., 1983. Successful storage of chilled rainbow trout (*Salmo gairdneri*) spermatozoa for up to 34 days. Aquaculture 31:269-274.

Stoss, J. and Refstie, T., 1983. Short-term storage and cryopreservation of milt from Atlantic salmon and sea trout. Aquaculture 30:229-236.

Szöllösi, D. and Billard, R., 1974. The micropyle of trout eggs and its reaction to different incubation media. J. Microsc. 21:55-62.

Takano, K., Hiroi, O., Yasukawa, M., and Suetake, T., 1973. Studies on the retention of gametes of salmonid fishes. 1. On the fertility of chum salmon eggs after storage. Sci. Rep. Hokkaido Salmon Hatchery 27:31-37.

Tomasik, L., 1974. Survival of "dry" salmonid sperm at temperatures above 0°C. Act. Ichthyol. Pisc. 4:4-8.

Truscott, B., Idler, D.R., Hoyle, R.J., and Freeman, H.C., 1968. Subzero preservation of Atlantic salmon sperm. J. Fish. Res. Board Can. 25:363-372.

Weil, C. and Crim, L.W., 1982. A study of methods of administering LHRH analogs: advancement of spermiation in the land-locked salmon (*Salmo salar*). In: C.J.J. Richter and H.J.Th. Goos (Comp.), Reproductive Physiology in Fish, PUDOC, Wageningen. pp. 26.

Weil, C. and Crim, L.W., 1983. Administration of LHRH analogues in various ways; effect on the advancement of spermiation in prespawning landlocked salmon *Salmo salar*. Aquaculture 35, (in press).

Winnicki, A. and Tomasik, L., 1976. Spermatocrit as a method for biological evaluation of fish sperm. Act. Ichthyol. Pisc. 6:3-7.

Withler, F.C. and Morley, R.B., 1968. Effects of chilled storage on viability of stored ova and sperm of sockeye and pink salmon. J. Fish. Res. Board Can. 25:2695-2699.

ABSTRACTS–Broodstock Management/ Husbandry

The Timing of Ovulation and Stripping and Their Effects on the Rates of Fertilization, Eyeing, Hatching and Swim-up in the Rainbow Trout, *(Salmo gairdneri)*

J. R. C. Springate, N. Bromage, and J. Elliott, University of Aston, Birmingham, United Kingdom

Britain's rainbow trout farmers with only 40% of their eggs surviving as 5g. fry, need 70-100 million eggs p.a. to satisfy the annual requirements of the table fish producers. Although the eggs are ovulated under conditions of intensive culture, they are not oviposited and remain in the body cavity until they are artificially stripped. During this time a process of ripening occurs and thus the timing of stripping in relation to ovulation is likely to have profound effects on the subsequent viability of the egg. Correct timing of stripping may enable significant improvements to be made in the number of eggs which can be fertilized.

In this study eight female rainbow trout, maintained in 10°C water, were examined daily until mature ova could be manually stripped from them (day 0). From this day onward approximately 100 eggs were stripped from each fish every other day for three weeks. Eggs from each stripping were subdivided for fertilization, eyeing, hatching and swim-up rate determinations, as well as for wet and dry weight and other analyses. Each batch of eggs was fertilized with the milt of two males. Blood samples were taken at each stripping for vitellogenin and steroid analyses.

The results show that the four developmental stages are closely correlated, and that poor fertilization rates are followed by reduced success at each of the subsequent developmental hurdles The timing of stripping and fertilization after the eggs have been released into the abdominal cavity is an all-important determinant of egg and fry survivals. Maximum productivity is shown to be achieved when the eggs are fertilized 5 days after they are ovulated. Ways in which this timing can be exploited commercially are discussed.

Development of an Arctic Charr *(Salvelinus alpinus)* Broodstock

M. H. Papst and G. E. Hopky, Department of Fisheries and Oceans, Winnipeg, Manitoba, Canada

The results of preliminary experiments in the development of a hatchery brood stock of arctic charr, *Salvelinus alpinus*, are described. The successful culture of arctic charr in a pilot commercial freshwater system and the difficulties in obtaining spawn from the wild have made the development of such a brood stock essential. The alternate use of warm (13°C) and cold (6.0°C) water, brought a group of charr to maturity in four years, compared to ten to eighteen years for natural stocks. Mature females were significantly smaller than those observed in the wild. Mature males were larger than females but smaller than mature males observed in the wild. The observed 1,769 eggs per female was lower than the mean 4,781 observed in the wild stock and the mean egg diameter of 4.9 mm was less than the 5.1 mm reported for the wild stock. The observed maturity index of 11.4% was only slightly less than the 12.8% reported for mature females in the natural stock. Mortality in eggs prior to eyeing was 75% in the cultured group, compared with 12% for the parental stock obtained as eggs from the wild and incubated under the same conditions. Time to hatching was similar for both groups. Examination of dead eggs revealed a significant number of unfertilized eggs in some of the spawned groups, while others exhibited the majority of mortalities in the cleavage and embryo stages. Survival and growth of eleutheroembryo, alevin and fry were comparable with those of stocks obtained from the wild. The occurrence of precocious maturity among intensively reared stocks was observed. Factors contributing to the poor survival of hatchery spawned arctic charr eggs and current culture experiments are discussed.

Measurement of a Sex-specific Protein in the Skin Mucus of Premature Coho Salmon *(Oncorhynchus kisutch)*

M. R. Gordon, T. Owen, and L. Hildebrand, B.C. Research, Vancouver, B.C.*
** Helix Biotech Ltd., Richmond, B.C., Canada*

The work reported here is the initial experiment we undertook to confirm our working hypothesis that a simple, rapid, non-interventive technique for sexing premature salmonids was possible without blood analysis. The final objective is to develop a "litmus paper" type of method which is harmless and unstressful and, most importantly, will determine the sex of fish several months before spawning.

Analysis of skin mucus specimens from fish sampled up to eight months prior to spawning revealed the presence of vitellogenin, a female-specific protein. The occurrence of this protein in the skin mucus of female fish was coincident with the initial development of the female gonads. These preliminary findings indicated that the concentration of vitellogenin in the mucus was greatly increased and readily discernible when female gonad weight was >1% and >6% of body weight, some 4–5 months prior to spawning. At this time no visible external signs of maturation was evident in male or female coho from this stock.

A practical method for sex determination that can be used by fish farmers, ranchers and hatchery biologists is currently under development.

Some Factors Affecting the Preservation of Salmonid Spermatozoa

A. W. Erdahl, D. A. Erdahl, E. F. Graham, University of Minnesota, St. Paul

Research was undertaken to study various factors influencing the storage of fresh (nonfrozen) and frozen semen of several salmonid species. The storage of fresh semen was investigated in regard to storing semen for use within one or two weeks from the time of collection. Frozen semen was investigated as a means of storing semen from year to year. Species studied included: chinook salmon *(Oncorhynchus tschawytscha)*, rainbow trout *(Salmo gairdneri)*, brown trout *(Salmo trutta fario)*, and brook trout *(Salvelinus fontinalis)*.

The diluent used in all studies was formulated according to Erdahl and Graham (1980). Non-frozen semen indicated a decreasing fertility rate with increasing storage time. Using chinook salmon semen at dilution ratios of less than 1:5 (1 part semen to 5 parts extender), fertility remained above 85% at 48 hour storage. Dilution ratios of 1:10 or above yielded fertility levels that decreased after 30 min. storage and were less than 2% fertile at 48 hours. Mean fertility values for brown trout and rainbow trout semen diluted 1:2 with extender decreased from 83 to 59% over 9 days.

Cyroprotective agents investigated included dimethyl sulfoxide (Me_2SO), ethylene glycol and glycerol. With all four species investigated the addition of Me_2SO to the extender in concentrations less than 7% (vol/vol) resulted in fertility approximating diluted semen without cryoagent (95%).Ethylene glycol resulted in fertility rates 20% below Me_2SO and glycerol showed highly detrimental effects on sperm cells even at very low concentrations.

Frozen semen of brown trout and brook trout diluted 1:2 with 7% Me_2SO as the cryoprotective agent resulted in an average fertility of 54% although the range was from 3 to 98% fertile dependent on trial. Semen frozen in the 0.25 cc straw resulted in fertility averaging 66% compared to semen frozen in the 0.10 cc pellet which averaged 10% fertile eggs over all trials conducted. Frozen semen stored at −79°C and −196°C showed no significant differences in fertility.

Cryopreservation of Milt and Its Application to Atlantic Salmon Farming

R. Alderson and A. J. MacNeil, Marine Harvest Ltd., Lochailort, Inverness-shire, United Kingdom

Recent studies by Stoss and others have shown the feasibility of long term storage of Atlantic salmon milt by cryopreservation. This technique could have value in the commercial farming of salmon by enabling milt of one year class to be used to fertilize eggs of the following year classes. This would improve the scope for rotational line crossing to avoid inbreeding depression in highly selected broodstock lines. The opportunity for using the technique will however be limited by the ease with which high levels of fertility can be achieved in full sized egg batches.

Initial trials using milt frozen in straws and batches of 300-400 eggs gave up to 90% of control fertilization with certain combinations of freezing conditions and cryoprotectants. Encouraged by these results, improvements were made in the method of milt collection which enabled large quantities of uncontaminated milt to be obtained. A further test with batches of 4,000 eggs, which were stripped and fertilized with cryopreserved milt on the sea cages holding the broodstock, gave fertilization rates of 30% and 70% of control levels. These results will be discussed further in the context of the difficulties which still remain for the wider use of cryopressured milt in the salmon farming business.

Successful Development of Pink Salmon Broodstock Husbandry Program and Facility Design at Port San Juan in Prince William Sound, Alaska

Brian Allee, Prince William Sound Regional Aquaculture Cooperative, Cordova, Alaska

The historical development (1975–1980) of the Port San Juan pink salmon hatchery was described by Kerns in 1980. The initial design involved broodstock being held in floating net pens in saltwater at salinities of 20–30 gms/l. This approach resulted in significant prespawning mortality and low egg viability.

Modifications each succeeding year have resulted in increased survival during maturation and improved egg viability. The present system incorporates volitional immigration from saltwater to freshwater during the maturation process. Data will be presented which summarizes maturation survival, egg viability and environmental parameters for eight specific brood years. Based upon this experience, a maturation program and facility design will be sponsored for a 200-million pink and 100-million chum salmon egg capacity hatchery in Prince William Sound.

Breeding Guidelines for Abating Inadvertent Genetic Change in Trout Hatchery Brood Stocks

S. R. Phelps, W. B. Schill, R. C. Simon, U.S. Fish and Wildlife Service, Kearneysville, West Virginia

The importance of maintaining genetic variation in brood stocks in becoming well recognized as a key factor for successful salmonid husbandry. However, our research indicates that the loss of genetic variation in brood stocks is often rapid, resulting in inbreeding levels correlated with a decrease in fish performance in only a few generations. This rapid loss of genetic variation is mainly inadvertent and is caused by genetic drift.

Inadvertent loss of genetic variation has to be minimized so that the available genetic variation can be used to produce selective gain and/or remain in the brood stock.

Recent recommendations of the numbers of brood stock fish to use for breeding to obtain an acceptable rate of genetic drift in brood stocks are based on idealized breeding conditions and are not directly applicable to fish culture operations. We have used population genetic data from trout hatchery brood stocks, test mating, and computer simulations to determine practical brood stock spawning procedures and developed guidelines compatible with fish culture operations to minimize inadvertent loss of genetic variation.

The number of breeding adults needed to maintain an acceptable rate of genetic change depends on the type of breeding methods used and the amount of genetic variation in the brood stock. We found two breeding methods, pair mating and an egg pooling procedure, to be effective in reducing the loss of genetic variation. Population genetic data from several trout strains have indicated genetic differences between various lots taken throughout the spawning season and also year to year variation. Thus, fish for the next brood stock generation have to be taken throughout the spawning cycle. Crossing generations will reduce year to year variability. By taking the necessary steps to minimize inadvertent genetic change in a brood stock, the fish culturist need not be concerned with inbreeding from mating closely related individuals.

The Relationship between Fertility of Rainbow Trout Eggs and Motility of Spermatozoa as Evaluated by Quasielastic Light Scattering Techniques

R. D. Moccia and K. R. Munkittrick, International Aquaculture Developments, Erin, Ontario; T. Craig and F. R. Hallet, University of Guelph, Canada

In many vertebrates, it is commonly assumed that spermatozoan motility is a reasonable indicator of fertility. Although there have been several investigations into the relationship between fertility and motility, few of these have utilized objective motility estimation techniques, and in general, results have been of low predictive value. It is possible that such a relationship may be obscured by the limitations of subjective motility estimation. Historically, subjective motility estimation is variable, not highly repeatable, and inaccurate when the sample contains either large numbers of dead sperm or sperm that are swimming abnormally.

Quasi-elastic light scattering (QELS) techniques offer repeatable, objective, quantitative evaluations of large numbers of sperm simultaneously. This study proposed to evaluate the relationship between fertility and motility in rainbow trout *(Salmo gairdneri)* using QELS techniques.

Gametes were collected from brood stock at a commercial trout hatchery. Semen was evaluated for spermatocrit, density, subjective motility, objective (QELS) motility and ability to fertilize pooled, replicated egg samples. A second trial late in the spawning season also involved cryopreservation of semen and fertilization of pooled egg samples.

There was a statistically significant relationship between sperm density and spermatocrit, although neither parameter could be related to subjective motility, objective motility or fertility. Objective estimates of motility were repeatable but needed standardization with respect to density and dilution. There was day-to-day variability in fertility and in the relationship between both subjective and objective motility estimates. Estimates of fertility were also variable. Cryopreserved samples showing the best subjective motility pre-freezing gave fertility, while those with poor pre-freezing motility gave none.

Developmental Rate, Fecundity, and Egg Size in Atlantic Salmon, *Salmo salar* L.

J. E. Thorpe, Department of Agriculture and Fisheries for Scotland, Pitlochry, Perthshire, United Kingdom

Egg number and egg size vary widely between individual spawners within stocks, and between stocks in Atlantic salmon. Early papers attempted to establish mean fecundity measures to characterize particular populations, or to supply rule-of-thumb estimating methods for fishery managers. More recently interest has been focussed, especially in the Soviet Union and Scandinavia, on temporal variation in size-fecundity relationships, and in age-fecundity relationships, within individual stocks. These studies are reviewed, and new data presented, evaluating the inter-relationships of developmental rate (river and sea age) on egg number and egg size in the River Almond salmon stock, Tayside, Scotland. In particular, it is shown that the rate of development of females during the juvenile riverine phase influences both the number and size of eggs produced subsequently. The general life-history strategy significance of this is discussed.

Factors Related to the Relatively Low Hatching Output in the Production of Salmon and Trout in Norwegian Commercial Hatcheries

Yngve Ulgenes and Gunnar Nævdal, Institute of Marine Research, Direktorate of Fisheries, Matredal, Norway

Until 1977-78 wild caught Atlantic salmon *(Salmo salar)* were commonly used for broodstock in the Norwegian fish farming industry. Later the demand for smolt increased rapidly and the use of reared brookstock became necessary for production of eggs for hatcheries. For experiments on genetic improvement, reared broodfish are also a necessity. As part of a program to improve the output from commercial hatcheries, the situation at several fish farms has been investigated. Problems associated with broodstock husbandry include the need to extend the spawning season, high mortality of broodfish during maturation and after spawning (stripping), and high mortality of the eggs during incubation. A great variation between year classes and between fish farms is experienced. The use of wild broodstock causes fewer problems than use of reared broodstock. Reasons for these problems and variations are still unclear but three factors seem to be involved: Stress on the fish in captivity, incomplete feed composition and incorrect feeding habits. Further handling of eggs during sensitive periods of incubation may be a fourth factor. In Western Norway, acid rain and toxic levels of aluminum in freshwater supplies are also problems.

Remote Location Hatchery Development

D. E. Pflug, W. Larrick, L. Fortier, R. W. Beck and Associates, Seattle*
**Southern Southeast Regional Aquaculture Association, Alaska*

As suitable sites for fish hatcheries become harder to find near population centers, salmon aquaculturists along the coasts of the Pacific Northwest, British Columbia and Alaska will begin to look toward more remote locations as increasingly viable alternatives for future construction. The development and successful operation of aquaculture facilities at remote locations calls for a critical evaluation of site characteristics and operational factors not normally considered in such detail for facilities in more populated areas. Two major considerations involve (1) developing a dependable and efficient power source for the hatchery and residence buildings, and (2) solving the logistical problems associated with transporting fish, hatchery supplies and personnel.

In addition to these major concerns, many other components of a hatchery project must be altered to accommodate remote siting problems. Economic, biological, engineering and construction considerations must all be accounted for and integrated into the overall development scheme to ensure a successful operation. The economics of becoming self-reliant often prove prohibitive unless innovative engineering methods are applied.

This paper provides some insight into problems associated with developing a remote salmonid hatchery facility. Development is viewed from the perspective of the biologist, the fishery manager and the engineer. A case study is examined and used to illustrate many of the points raised. The author emphasizes the importance of thorough planning and the need to integrate engineers and biologists into a single team to produce a facility capable of operating successfully over a wide range of natural and institutional conditions.

Roundtable Discussions

Discussion Leaders

Brian Allee
Prince William Sound Regional
Aquaculture Corporation
Cordova, AK

Rolland Billard
Laboratoire de Physiologie de Poisson
Institut National de la Recherche Agronomique
Jouy-en-Josas, France

Edward Brauer
DomSea Farms, Inc.
Bremerton, WA

L.W. Crim
Memorial University of Newfoundland
St. John's, Canada

Walton W. Dickhoff
University of Washington
Seattle, WA

Graham Gall
University of California-Davis
Davis, CA

Ronald W. Hardy
University of Washington and
National Marine Fisheries Service
Seattle, WA

Lee Harrell
National Marine Fisheries Service
Manchester, WA

William K. Hershberger
University of Washington
Seattle, WA

David Higgs
Department of Fisheries and Oceans
West Vancouver, B.C., Canada

Bill Hopley
Washington Department of Fisheries
Olympia, WA

George A. Hunter
Department of Fisheries and Oceans
West Vancouver, B.C., Canada

Robert N. Iwamoto
University of Washington
Seattle, WA

James Lannan
Oregon State University
Newport, OR

Conrad Mahnken
National Marine Fisheries Service
Manchester, WA

William McNeil
Oregon Aqua Foods
Springfield, OR

Stacia Sower
University of New Hampshire
Durham, NH

Gary Thorgaard
Washington State University
Pullman, WA

J.E. Thorpe
Department of Agriculture
and Fisheries for Scotland
Pitlochry, Perthshire, Scotland

Fred Utter
National Marine Fisheries Service
Seattle, WA

Roundtable Discussions

Stacia A. Sower,[1] and Robert N. Iwamoto,[2].

The roundtable discussions were directed toward elaborating the benefits and limitations of applying current research results to salmonid production. Additionally, aspects of salmonid reproduction requiring further research and definition were identified. Most discussions focused on sex control, induced ovulation, nutrition, and chromosomal manipulations; little was said about environmental manipulations.

In the sections below, the questions addressed are followed by a summary of discussions at twelve different tables by ten to fifteen participants at each table.

Induced Ovulation, Sex Control, And Environmental Manipulations

What are the effects of these manipulations on the growth, physiological development, migratory behavior, and reproductive status of resultant progeny? Are one or a few generations of monitoring effects sufficient justification for production-scale application? Should criteria for production-scale application be identical for captive and ocean-going broodstocks?

Induced Ovulation: Induced ovulation appeared to be the most promising and least controversial technique discussed. Often when adult salmon return to freshwater hatcheries, they do not spawn before they die. In other cases, the reproductive organs do not fully mature before the fish die. Accelerated maturation/ovulation helps overcome these problems. Researchers attending this symposium have demonstrated that maturation/ovulation can be accelerated with low doses of a combination of an analogue of the brain hormone (gonadotropin-releasing hormone, GnRH) and a pituitary gland hormone (gonadotropin), or two injections of GnRH. As adults, treated salmon will spawn earlier than normal and release their eggs before they die.

One stated benefit of induced ovulation is that aquaculture companies are able to take more eggs and thereby reduce their dependence on eggs supplied by public hatcheries. Another benefit is that eggs can be obtained earlier, which means a quicker start for the freshwater growth of young salmon. This is especially advantageous in hatcheries that have cold water (which slows the growth of young salmon) rearing facilities and in hatcheries that produce one-year smolts. Recently, these treatments have been used on a production scale in one

1 . University of New Hampshire
2 . University of Washington

private salmon aquaculture operation at a cost effective dose of approximately 3 cents per fish.

In discussions about induced ovulation, two major problems emerged. The first was that only one study has been conducted to assess the survival and fitness of subsequent generations of treated fish, and no studies have been done to determine hormone effects on reproductive performance of the progeny of treated fish. The second was that the U.S. Food and Drug Administration has not approved the hormone treatment, and so far as is known there have been no clearance studies on salmon treated with hormones. Until the progeny testing is completed and FDA approval occurs, these hormones (which are proteins that break down within hours after injection) cannot be used on a regular production basis.

One further problem, though not so serious, concerns the use of hormonal treatment at inappropriate times, e.g., too early. If salmon are injected within approximately 6 weeks of the time of normal ovulation, the fish will ovulate but egg survival will be low. The proper dosage and timing of treatment needs to be determined for each salmonid species and strain, since maturational times vary.

Sex Control: Sterilization of fish and production of all-female groups by early steroid treatment were the most controversial techniques discussed.

R. Alderson of Marine Harvest, Ltd., U.K., emphasized that the public may not be receptive to the idea of eating trout/salmon treated with steroids. Marine Harvest avoids the use of steroids for this reason. However, it was pointed out that clearance time for steroids used in treatments is 30-60 days, so steroid residual may not be a problem since salmon are not consumed until they are adults (age 2-4 years).

Sex reversal from male to female by feeding and/or by immersion in estradiol is still in the experimental stage. Doses of estradiol must be carefully monitored since it has been shown that high levels of estradiol can induce mortalities.

Another method of inducing sex reversal was discussed by J. Springate of the University of Aston, U.K. He commented on a British market taste preference for female rather than male trout. As a result, some trout farms currently produce all-female fish by first treating fry (males and females) with methyltestosterone which induces genotypic females to produce sperm. These genotypic females/phenotypic males are then crossed with normal females which results in all-female progeny. Because it is the second generation that is used for production, concerns about steroids in fish that will be consumed by the public are eliminated. However, further experiments were deemed necessary if the progeny of treated parents are to be used as broodstock—it would have to be shown that the reproductive development in the progeny is normal before use in production.

Although there was not much support from other participants, E. M. Donaldson and G. A. Hunter, Fisheries and Oceans Canada, advocated the use of hormonally sterilized salmon in the fisheries. Donaldson's lab released an experimental group of coho salmon that had been sterilized as fry with methyltestosterone approximately 1 4 years prior to release. Their objectives were to eliminate the production of coho jacks (small, early spawning males) and to produce larger fish for the commercial and sports fisheries. An assumption of the experiment is that sterile males do not return to the hatchery and thus remain in the ocean. Nevertheless, a few coho salmon subjected to the sterilization treatment did return as production males.

Participants questioned the cost-effectiveness of this technique since it sterilizes females as well as males and thus eliminates a valuable source of eggs. Others wondered about its effectiveness since some scientists have shown that doses similar to those administered have not completely sterilized the treated fish. Still others expressed concern about steroid effects on the migratory behavior of salmon and wondered whether these fish might stray into other streams and attempt to spawn with wild salmon. Finally, participants debated the effects of steroids on overall fitness. They pointed out that androgens—which are hormones important for growth and are normally produced by the gonads—would not be produced. Therefore they wondered how well steroid-treated fish would grow and survive in the marine environment.

It was generally agreed that even if these kinds of manipulations prove effective, for the time being these techniques should be applied to captive broodstocks only. The consensus was that steroids affect processes in fish other than reproductive ones and that steroid influence on these systems (osmoregulation, behavior, smoltification) should be studied thoroughly before use in a sea-ranching operation.

Environmental Manipulations: This topic did not stimulate much discussion. In general, manipulations of photoperiod and temperature are feasible only where trout or salmon are reared entirely in captivity. Photoperiod manipulations by Europeans have produced trout eggs at times other than normal. This could be especially valuable in England where more than 30 million eggs are imported annually and egg production throughout the year is desirable.

Participants were concerned that fish released as smolts at times other than "normal" might have adverse impacts on the ecosystem. They also wondered if desmoltification would occur if coho salmon, not fully "smolted", were released into seawater.

All agreed that more information is needed about the effects of environmental factors singly and in combination with one another. This would help producers and policy makers decide what factors could be manipulated to improve salmon and trout stocks.

Nutrition

Diet is a crucial aspect of salmonid broodstock production, yet the dietary requirements for larger, maturing salmonids are not fully known. What are the effects of different diets on maturation, development, disease and growth? How do dietary deficiencies affect other physiological or endocrinological research results?

The major conclusion in the discussions related to nutrition was that there is a pressing and immediate need for more work and research in broodstock nutrition. Little is known about nutrient requirements of salmonids at different stages in life history, and it is especially important to learn these requirements for salmon in the marine environment and for broodstock.

Although certain diets are adequate in some situations but not in others, trout farmers present indicated that that this is not a serious problem. However, they did say that variable quality of fish feed is a problem. Some said that feed manufacturers pay a great deal of attention to starter and broodstock diets–however, the fish in between get "junk food." They speculated that the quality of fish feeds varies with the availability of ingredients, e.g., fish meal. Variable diets were also suggested as a factor that could skew research results.

Participants agreed that increased knowledge of nutrient requirements and appropriate feeding strategies plus improved feed quality would produce better quality trout, and ultimately, improve egg production. To gain this knowledge, long-term studies of broodstock nutrition were deemed essential.

Genetics and Reproduction

Gamete Banks. Have we sufficiently demonstrated that gamete banks are an effective means of preserving the genetic uniqueness of endangered strains? How can we be reasonably assured that genome changes in captive broodstock supplying progeny for enhancement are minimized?

Chromosomal Manipulation. What are the implications of co-mingling polyploid individuals and gynogenetic females with "normal" diploid fish?

It has been shown that heat shock and hydrostatic pressure treatments of salmonid eggs induce triploid individuals having three sets of chromosomes as contrasted with normal diploids having two sets of chromosomes. During sexual maturation, the growth rates of triploids appear greater than growth rates of normal diploid individuals. There are, however, questions about whether growth rates of triploid fry and juveniles equal those of normal diploid fish, and about the saltwater adaptability and the survival rates of chromosomally manipulated individuals. Definitive conclusions cannot be made until the results from tests under controlled conditions are available.

There was considerable interest in the application of triploidy to produce species hybrids with desirable characteristics of one or both parental species (e.g., disease resistance or early saltwater tolerance) when a diploid hybrid would not be viable. At present, this type of "genetic engineering" is considerably more feasible than the direct gene transfers (DNA hybridization) investigated for other species.

The potential co-mingling of triploid fish and gynogenetic females with normal diploid fish was not viewed as a critical issue. Although most experiments have been strictly exploratory, they appear to have wide application with captive fish. On the technical side, considerable variability (rate of conversion) in the triploid individuals produced indicates that the state of maturity of the female and/or the quality of eggs prior to heat or pressure shock may affect significantly the technique's success.

Another area of discussion concerned the cryopreservation of gametes (principally milt) which could be a useful tool for maintaining genetic variability in endangered stocks (gene banks), and of potentially greater application for broodstock management and selection programs. Attempts have been made to adapt the technique to production scale programs, but there are questions about the correct timing of the process and the quality of sperm after long-term storage.

There was almost universal agreement that "something is lost" during the domestication of wild strains. An intriguing, yet unanswered, question that was raised was the degree to which the domestication process can be reversed as in feral livestock.

Overall, there was a great deal of interest in selective breeding. Participants questioned whether advances in the breeding of salmonids could be as significant as those made in the breeding of plants and agricultural animals. It was emphasized that salmonid traits cannot be so easily selected because stocks have not been well-defined and the genetic basis of particular traits has not been well-established. Selection and crossbreeding can proceed on a rational basis only after individuals have been identified and marked and performance monitored for several generations. The development of a marker that could be used to identify individual fish would advance the genetic selection process significantly.

EPILOGUE

The papers presented and discussions which ensued were presented almost entirely by scientists from governments, universities, and commercial organizations. We believe this symposium, designed for exchange of information, exceeded its goal of evaluating fully the state of the art and of describing and assessing current commercial applications. Realization of the need for certain areas of research was heightened. Nutrition is a good example. This area of inquiry may provide the key to successful salmonid aquaculture, yet little is known about nutritional requirements of reproductively active salmon. Advances in knowledge about nutrition could be applied immediately.

In other subject areas of the meetings, it was apparent to the fish culturists and scientists that many aspects of endocrinology and genetics presented are not yet ready for commercial application. The potential effects of techniques which may appear promising cannot be fully known or appreciated without complete information that is possible only through basic research. Moreover, endocrine, genetic, or environmental manipulations in trout farms and captive broodstock programs must be viewed differently from those in ocean-ranching programs. For example, selection pressures on released fish are certainly different from those on farm-bred fish. Sex reversal or sterility induced by steroid hormones may have greater consequences for the overall fitness and long-term survival of salmonids released into the wild than for those maintained in captivity. There was overwhelming agreement among the participants that adequate studies must be done and the results fully evaluated before commercial applications, however promising, are made.

The age at which salmonids mature makes long-term studies difficult. However, such studies are essential and require the long-term commitments of researchers and their sponsors. In short, cautious evaluation and conservative approaches should characterize any use of enhancement techniques for salmon.

There are many processes involved in salmonid reproduction which are not fully understood, yet are critical to the successful maintenance of salmonids as an important biological, economic, and cultural resource. The information garnered from this symposium and the enthusiasm generated by its participants have contributed to an optimistic appraisal of future prospects. As Bill McNeil observed in his keynote address: ''Collaboration, communication, open minds, and effort to understand could inevitably save a resource and open avenues for cooperation.''

The Aquaculture Data Base

Carol Rideout[1]

The Aquaculture Data Base is a centralized source of bibliographic information for literature relevant to aquaculture. All species cultured in water, including salmonids, are covered. The data base is publicly available as File 112 of the DIALOG Information Retrieval System. Custom bibliographies are easily produced via direct, online searches or by assisted searches. Microfiche or paper copies of non-copyrighted articles from the bibliography may be obtained from the Virginia Insititute of Marine Science. Copyright holders have granted permission for many of the other articles to be copied as well. Literature is collected from a wide variety of sources including professional journals and newsletters, proceedings, books, government publications, research reports and unpublished papers. Data base holdings are continually expanded at a rate of 100 new entries per month. The Aquaculture Data Base is funded through a cooperative agreement among the National Agriculture Library, National Oceanic and Atmospheric Administration, and the Virginia Institute of Marine Science.

The number of unique, useful salmonid bibliographies that may be created through searches of the Aquaculture Data Base is practically unlimited. By August 1983, 310 of approximately 10,000 documents described in the data base concerned the culture of coho salmon, *Oncorhynchus kisutch*; 221 were about chinook salmon, *Oncorhynchus tshawytscha*; 109 about sockeye salmon, *Oncorhynchus nerka*, 197 on Atlantic salmon, *Salmo salar* and 743 concerned culture of the trout, *Salmo gairdneri*.

Three searches of the data base were conducted as examples of its potential as a source of salmonid reproduction information. A search was designed to produce a custom bibliography about induced spawning and artificial fertilization of *Oncorhynchus* species. The genus name, "*Oncorhynchus*," was cross-referenced with the terms "induced spawning" or "artificial fertilization." The resulting bibliography listed 25 relevant articles. Another search of the Aquaculture Data Base was designed to find literature published or produced 1977–1982 concerning selective breeding and brood stock management of *Salmo* species. A search for the terms "brood stock" or "selective breeding" with the genus name "*Salmo*" and with publication dates "1977" through "1982" located 45 references. A third search for salmonid genetic studies resulted in 38 references.

[1] Virginia Institute of Marine Science, Gloucester Point, VA

Induced Spawning and Artificial Fertilization of *Oncorhynchus* Species

Anon., Different Methods of Fertilizing Sockeye Salmon Eggs Studied at Bristol Bay Hatchery. Northwest Fisheries Center Monthly Report 1975 (November), 18-19. (Unpublished material)
Descriptors: artificial fertilization; sockeye salmon; egg; hatchery; fish
Genus Species: Oncorhynchus nerka

Bailey, J.E., Pella, J.J., Taylor, S.G., 1977. Effects of Substrate Depth, Seeding Density, and Water Flow on Production of Pink Salmon Fry from Incubators Using Plastic Turf. Northwest and Alaska Fisheries Center Processed Report, (February), 44 pp.
Descriptors: pink salmon; incubation; Alaska, production; hatchery; fry; substrate; seed; stock density; survival; flow; equipment; collecting; fertilization; artificial fertilization; temperature; incubator; length; weight; mortality; anadromous fish
Genus Species: Oncorhynchus gorbuscha

Bams, R.A., 1970. Evaluation of a Revised Hatchery Method Tested on Pink and Chum Salmon Fry. Journal of the Fisheries Research Board of Canada, 27(8):1429-1452.
Descriptors: fry; hatchery; pink salmon; chum salmon; anadromous fish; incubation; growth; survival; fresh water; gravel; development; artificial fertilization; artificial spawning; migration; temperature; length; weight; egg
Genus Species: Oncorhynchus keta; Oncorhynchus gorbuscha

Borgese, E.M., 1979. A Net of Connected Meshes (Sea Farming Round the World). Oceans 12(2):9-14.
Descriptors: mollusk; oyster; India; rope culture; seaweed; tray culture; growth; labor; mussel; spat; lobster; crustacean; larva; fry; market; induced spawning; Thailand; eyestalk; pituitary; shrimp; hatchery; milkfish; marine fish; stock density; Philippines; production; anadromous fish; adaptability; coho salmon; green turtle; reptile; baitfish
Genus Species: Chanos chanos; Oncorhynchus kisutch

Brown, E.E., 1977. World Fish Farming: Cultivation and Economics. 1. United States of America. World Fish Farming: Cultivation and Economics, The AVI Publishing Company, Inc., pp. 4-71.
Descriptors: fresh water fish; rainbow trout; United States; fee fishing; feed composition; history; productivity; temperature; management; incubation; economics; fry; automatic device; feeder; survival; hatching; channel catfish; pond culture; brood stock; feeding; egg; life history; induced spawning; predation; water quality; harvesting; production; crop rotation; land use; recreation; food conversion; raceway culture; market; bullhead catfish; American eel; catadromous fish; fresh water prawn; largemouth bass; crayfish; crustacean; predator control; trap; equipment; anadromous fish; chum salmon; coho salmon; chinook salmon; Abernathy diet; cage culture; walleye, northern pike; sockeye salmon; brook trout; brown shrimp; embayment culture; pink shrimp; white shrimp; striped bass; cutthroat trout; muskellunge; bluegill; lake trout
Genus Species: Salvelinus namaycush; Salvelinus fontinalis; Salmo trutta; Salmo gairdneri; Salmo clarki; Ictalurus punctatus; Anguilla rostrata; Macrobrachium rosenbergii; Oncorhynchus keta; Oncorhynchus kitsutch; Oncorhynchus tshawytscha; Oncorhynchus nerka; Penaeus setiferus; Penaeus aztecus; Penaeus duorarum; Stizostedion vitreum; Esox lucius; Micropterus salmoides; Lepomis macrochirus; Roccus saxatilis; Esox masquinonongy

Dickhoff, W., Sower, S., 1981. Hormone Research Benefits Salmon Aquaculture. University of Washington Sea Grant, Current Marine Research & Activities--News Leads, 1981, 1 pg.
Descriptors: anadromous fish; hormone; thyroid; blood; coho salmon; open water culture; hatchery; induced spawning; pituitary; juvenile; research; Washington
Genus Species: Oncorhynchus kisutch

Donaldson, E.M., Hunter, G.A., Dye, H.M., 1981. Induced Ovulation in Coho Salmon(*Oncorhynchus kisutch*). III. Preliminary Study on the Use of the Antestrogen Tamoxifen. Aquaculture, 1981, 26(1,2):143-154.
Descriptors: coho salmon; anadromous fish; ovulation; induced spawning; injection; survival; hormone
Genus Species: Oncorhynchus kisutch

Donaldson, E.M., Hunter, G.A., Dye, H.M., 1981. Induced Ovulation in Coho Salmon(*Oncorhynchus kisutch*). II. Preliminary Study of the Use of LH-RH and Two High Potency LH-RH Analogues. Aquaculture, 26(1,2):129-141.
Descriptors: hormone; induced spawning; ovulation; coho salmon; anadromous fish; injection; hatching; egg; survival
Genus Species: Oncorhynchus kisutch

Hjul, P., Editor, 1980. Alaska's Ranchers. Fishermen Join Fish Farmers to Boost Pacific Salmon Runs. Fish Farming International 7(4):4-5.
Descriptors: Alaska; anadromous fish; open water culture; restoration; artificial fertilization; financing; hatchery; pink salmon; chum salmon
Genus Species: Oncorhynchus gorbuscha; Oncorhynchus keta

Hobart, W., Editor, 1981. Hormone Research Aids Pacific Salmon Culture. Marine Fisheries Review 43(9):25.
Descriptors: Hormone; anadromous fish; coho salmon; thyroid; open water culture; cage culture; hatchery; induced spawning
Genus Species: Oncorhynchus kisutch

Hough, S., Editor, 1978. Special Report. Aquaculture, the State of the Art. Calypso Log 5(2), Supplement 6p.
Descriptors: Aquaculture; history; problems; Asia; China; India; Indonesia; Japan; Philippines; Africa; Central America; United States; fresh water fish; mixed culture; grass carp; bighead carp; silver carp; nutrient; water hyacinth; rainbow trout; Idaho; catfish; production; feed composition; soy meal; fish meal; ecology; marine fish; milkfish; mullet; algae; sewage; pond culture; lobster; crustacean; brine shrimp; economics; cost; shrimp; research; oyster; mollusk; France; induced spawning; temperature; hatchery; fecundity; predator control; effluent; sewage treatment; flounder; flatfish; abalone; waste water aquaculture; pollution; salmon; anadromous fish; enclosure; legal aspect; herring; mussel; Spain; protein
Genus Species: Oncorhynchus; Chanos chanos; Ctenopharyngodon idellus; Hypophthalmichthys nobilis; Hypophthalmichthys molitrix

Huet, M., 1973. Breeding and Cultivation of Salmonids or Fish Culture in Cold Water. Textbook of Fish Culture. Breeding and Cultivation of Fish, pp. 59-110, Fishing News (Books) Ltd., London.
Descriptors: Cold water fish; water requirement; site selection; water quality; water volume; water supply; aquafarm; brown trout; food requirement; rainbow trout; Atlantic salmon; Danube salmon; Arctic char; rainbow trout; cutthroat trout; pink salmon; chum salmon; coho salmon; sockeye salmon; chinook salmon; Masu salmon; brook trout; habitat; distribution; lake trout; brood stock; sex ratio; sperm viability; egg viability; age; selective breeding; artificial fertilization; spawning time; holding artificial spawning; maturity; standard procedure; incubation; hatching; hatchery equipment; age; pond culture; development; fry; artificial food
Genus Species: Salmo trutta; Salmo salar; Hucho hucho; Salvelinus alpinus; Salmo gairdneri; Salmo clarki; Oncorhynchus gorbuscha; Oncorhynchus keta; Oncorhynchus kisutch; Oncorhynchus nerka; Oncorhynchus tshawytscha; Oncorhynchus masou; Salvelinus fontinalis; Salvelinus namaycush

Hunter, G.A., Donaldson, E.M., Dye, H.M., 1981. Induced Ovulation in Coho Salmon (*Oncorhynchus kisutch*). I. Further Studies on the use of Salmon Pituitary Preparations. Aquaculture 26(1,2):117-127.
Descriptors: Induced spawning; ovulation; coho salmon; anadromous fish; pituitary; gonadotropin; hatching; egg; survival; injection
Genus Species: Oncorhynchus kisutch

Hunter, G.A., Donaldson, E.M., Stone, E.T., Dye, H.M., 1978. Induced Ovulation of Female Chinook Salmon (*Oncorhynchus tshawytscha*) at a Production Hatchery. Aquaculture 15(2):99-112.
Descriptors: Ovulation; chinook salmon; anadromous fish; hatchery; mortality; induced spawning; British Columbia; injection; egg; size
Genus Species: Oncorhynchus tshawytscha

Jalabert, B., Goetz, F.W., Breton, B., Fostier, A., Donaldson, E., 1978. Precocious Induction of Oocyte Maturation and Ovulation in Coho Salmon, Oncorhynchus kisutch. Journal of the Fisheries Research Board of Canada 35(11):1423-1429.

Descriptors: Induced spawning; ovulation; oocyte; development; coho salmon; anadromous fish; gonadotropin
Genus Species: Oncorhynchus kisutch

Kinne, O., 1977. 5. Cultivation of Animals. 5 1 Research Cultivation (13) Pisces. Marine Ecology. A Comprehensive, Integrated Treatise on Life in Oceans and Coastal Waters, 1977, Vol. III, Part 2, pp. 968-1035.
Descriptors: Larva; fertilization; incubation; hatching; mortality; nutrition; ecology; feed composition; live food; diet; plaice; flatfish; marine fish; food density; haddock; northern anchovy; food selection; starvation; behavior; herring; antibiotic; Omaka; juvenile; rabbitfish; pellet; artificial food; Oregon moist pellet; salmon; anadromous fish; protein; amino acid; rainbow trout; fresh water fish; sole; mineral; vitamin; feeding; induced spawning; artificial spawning; sperm; puffer; striped mullet; research
Genus Species: Pleuonectes platessa; Melanogrammus aeglefinus; Engraulis mordax; Clupea harengus; Caranx mate; Siganus canaliculatus; Salmo gairdneri; Solea solea; Oncorhynchus; Fugu rubripes; Mugil cephalus

MacQuarrie, D.W., Vanstone, W.E., Market, J.R., 1979. Photoperiod Induced Off-Season Spawning of Pink Salmon (*Oncorhynchus gorbuscha*). Aquaculture 18(4):289-302.
Descriptors: Photoperiodism; season; pink salmon; anadromous fish; induced spawning; light; fecundity; Canada; brood stock; maturity; mortality; aquarium culture; trough culture
Genus Species: Oncorhynchus gorbuscha

McNeil, W.J., 1975. Perspectives on Ocean Ranching of Pacific Salmon. Proceedings of the Sixth Annual Workshop World Mariculture Society, Seattle, Washington, pp. 299-308.
Descriptors: Open water culture; salmon; anadromous fish; ecology; behavior; genetics; growth; survival; cost; juvenile; legal aspect; property rights; economics; social aspect; pink salmon; chum salmon; chinook salmon; sockeye salmon; incubation; feeding; gravel; equipment; induced spawning
Genus Species: Oncorhynchus gorbuscha; Oncorhychus keta; Oncorhynchus tshawytscha; Oncorhynchus kisutch; Oncorhynchus nerka

Mihara, T., Sano, S., Eguchi, H., Artificial Propagation of Salmon in Japan. Japan Fisheries Resource Conservation Association, Tokyo. Mariculture Series 5 (English translation from Japanese by Language Services Branch NMFS, NOAA, 20p.), 9999.
Descriptors: Japan; anadromous fish; migration; collecting; pond culture; site selection; water quality; oxygen; mortality; sockeye salmon; Masu salmon; artificial fertilization; egg; temperature; hatching; disinfection; hatchery; fry; length; homing; pink salmon
Genus Species: Oncorhynchus keta; Oncorhynchus masou; Oncorhynchus gorbuscha

Pillay, T.V.R. (Editor), 1972. From Research Institutions. FAO Aquaculture Bulletin 4(2):3-8, FAO 20669.
Descriptors: Mixed culture; eastern oyster; storage; pituitary preparation; catfish; grass carp; giant freshwater prawn; crustacean; shellfish; fertilizer; pond management; ascorbic acid; nutrient requirement; catfish; vitamin deficiency; dietary additive; fishery waste; feed composition; demand feeding; aquatic weed control; chemcontrol; raceway culture; coho salmon; shrimp; tuna; whitefish; mollusk; fish; mullet; induced spawning
Genus Species: Neothunnus macropterus; Thunnus thynnus; Penaeus aztecus; Oncorhynchus kisutch; Penaeus merguiensis; Macrobrachium rosenbergii; Clarias lazera; Ctenopharyngodon idellus

Pillay, T.V.R. (Editor), 1971. From Research Institutions. FAO Aquaculture Bulletin 3(3):2-7, FAO 17209.
Descriptors: Striped mullet; artificial spawning; induced spawning; photoperiodism; pond culture; Columbia; sabaleta; walking catfish; Mahseer; carp; fish; shrimp; crustacean; shellfish; growth rate; artificial food; Chinese carp; chinook salmon; rotenone; mixed culture; duck; thermal pollution; sublethal dosage; power station effluent; Poland; waste water aquaculture; eutrophication; crucian carp
Genus Species: Carassius carassius; Oncorhynchus tshawytscha; Penaeus setiferus; Penaeus aztecus; Brycon henni; Mugil cephalus

Refstie, T., Stoss, J., Donaldson, E.M., 1982. Production of All-Female Coho Salmon (*Oncorhynchus kisutch*) by Diploid Gynogenesis Using Irradiated Sperm and Cold Shock. Aquaculture 29(1,2):67-82.

Descriptors: Monosex culture; coho salmon; anadromous fish; gynogenesis; irradiation; sperm; brackish water; thermal shock; cold water; induced spawning; mortality; chromosome; British Columbia; Canada
Genus Species: Oncorhynchus kisutch

Smoker, W.W., Kerns, C.L., 1978. Artificial Salmon Spawning--A Manual. University of Alaska Sea Grant, AKU-H-78-001, Marine Advisory Bulletin No. 7, 1-21.
Descriptors: Artificial spawning; anadromous fish; collecting; holding; transport; cage culture; sex; chum salmon; pink salmon; egg; artificial fertilization; hatchery; handling
Genus Species: Oncorhynchus keta

Sower, S.A., Schreck, C.B., 1982. Sexual Maturation of Coho Salmon (*Oncorhynchus kisutch*): Accelerated Ovulation and Circulating Steriod Hormone and Ion Levels of Salmon in Freshwater and Seawater. Proceedings of the North Pacific Aquaculture Symposium, Anchorage, Alaska, August 18-21, 1980; Newport, Oregon, August 25-27, 1980. Alaska Sea Grant Report 82-2, 227-235.
Descriptors: Development; coho salmon; anadromous fish; hormone; fresh water; sea water; ovulation; induced spawning; steroid osmoregulation; mortality; reproduction
Genus Species: Oncorhynchus kisutch

Zirges, M.H.; Curtis, L.D., 1972. Viability of Fall Chinook Salmon Eggs Spawned and Fertilized 24 Hours After Death of Female. The Progressive Fish-Culturist 34(4):190.
Descriptors: Chinook salmon; fish; egg; artificial spawning; artificial fertilization; survival
Genus Species: Oncorhynchus tshawytscha

Selective Breeding and Brood Stock Management of Salmo Species, 1977-1982

Allendorf, F.W., Phelps, S.R., 1980. Loss of Genetic Variation in a Hatchery Stock of Cutthroat Trout. Transactions of the American Fisheries Society 109(5):537-543.
Descriptors: Genetics; hatchery; cutthroat trout; fresh water fish; brood stock; Montana
Genus Species: Salmo clarki

Avault, J.W., Jr., 1980. Salt—Useful Tool in Aquaculture. Aquaculture Magazine 6(3):40-41.
Descriptors: Salt; disease control; incubation; survival; malachitegreen; parasite; egg; saprolegnia; fungus; salmon; anadromous fish, channel catfish; delayed spawning; oyster; predator control; prawn; crustacean; growth; rainbow trout; brood stock; age; fresh water fish
Genus Species: Ictalurus punctatus; Salmo gairdneri

Bailey, J.K., Saunders, R.L., Buzeta, M.I., 1980. Influence of Parental Smolt Ageand Sea Age on Growth and Smolting of Hatchery-Reared Atlantic Salmon(*Salmo salar*). Canadian Journal of Fisheries and Aquatic Sciences 37(10):1379-1386.
Descriptors: Age; smolt; growth; hatchery; Atlantic salmon; anadromous fish; length; genetics; maturity; Canada; brood stock; sex
Genus Species: Salmo salar

Bergot, P., Blanc, J.M., Escaffre, A.M., Poisson, H., 1981. Effect of Selecting Sires According to their Number of Pyloric Caeca Upon the Growth of Offspring in Rainbow Trout (*Salmo gairdneri* Richardson). Aquaculture 25(2,3):207-215.
Descriptors: France; selective breeding; growth; genetics; size; rainbow trout; fresh water fish
Genus Species: Salmo gairdneri

Blanc, J.M., Chevassus, B., 1982. Interspecific Hybridization of Salmonid Fish. II. Survival and Growth Up to the 4th Month After Hatching in F_1 Generation Hybrids. Aquaculture 29(3,4):383-387.
Descriptors: Hybrid; survival; growth; anadromous fish; coho salmon; chinook salmon; brown trout; brook trout; rainbow trout; fresh waterfish; France; selective breeding; hatching; trough culture; Atlantic salmon
Genus Species: Oncorhynchus tshawytscha; Oncorhynchus kisutch; Salmo gairdneri; Salmo trutta; Salmo salar; Salvelinus fontinalis

Bowerman, M. (Editor), 1980. Science and Hard Work Build Trout Business. Australian Fisheries 39(6):10-11.

Descriptors: United Kingdom; fry; production; export; commercial firm; economics; rainbow trout; fresh water fish; selective breeding; hatchery; pond culture; raceway culture
Genus Species: Salmo gairdneri

Brezosky, P.E., Thoits, C.F. III, 1978. Operation and Maintenance of the Milford Hatchery: Performance Report July 1, 1977—June 30, 1978. New Hampshire Project No. AFC-4-1, 6 p.
Descriptors: Hatchery; management; anadromous fish; brood stock; coho salmon; Atlantic salmon; chinook salmon; hybrid
Genus Species: Oncorhynchus kisutch; Salmo salar; Oncorhynchus tshawytscha

Brezosky, P.E., Thoits, C.F. III, 1977. Operation and Maintenance of the Milford Hatchery. New Hampshire Project No. AFC-3, 22 p., Project Completion Report.
Descriptors: Hatchery; management; anadromous fish; coho salmon; hatching; New Hampshire; weight; feeding; growth; mortality; collecting; production; stocking; Atlantic salmon; survival; selective breeding; incubation
Genus Species: Oncorhynchus kisutch; Salmo salar

Brown, E.E., 1977. World Fish Farming: Cultivation and Economics. 1. United States of America. World Fish Farming: Cultivation and Economics, The AVI Publishing Company, Inc., pp. 4-71.
Descriptors: Fresh water fish; rainbow trout; United States; fee fishing; feed composition; history; productivity; temperature; management; incubation; economics; fry; automatic device; feeder; survival; hatching; channel catfish; pond culture; brood stock; feeding; egg; life history; induced spawning; predation; water quality; harvesting; production; crop rotation; land use; recreation; food conversion; raceway culture; market; bullhead catfish; American eel; catadromous fish; fresh water prawn; largemouth bass; crayfish; crustacean; predator control; trap; equipment; anadromous fish; chum salmon; coho salmon; chinook salmon; Abernathy diet; cage culture; walleye; northern pike; sockeye salmon; brook trout; brown shrimp; embayment culture; pink shrimp; white shrimp; striped bass; cutthroat trout; muskellunge; bluegill; lake trout
Genus Species: Salvelinus namaycush; Salevelinus fontinalis; Salmotrutta; Salmo gairdneri; Salmo clarki; Ictalurus punctatus; Anguillarostrata; Macrobrachium rosenbergii; Oncorhynchus keta; Oncorhynchus kisutch; Oncorhynchus tshawytscha; Oncorhynchus nerka; Penaeus setiferus; Penaeus aztecus; Penaeus duorarum; Stizostedion vitreum; Esox lucius; Micropterus salmoides; Lepomis macrochirus; Roccus saxatilis; Esox masquinongy

Busack, C.A., Gall, G.A.E., 1980. Ancestry of Artifically Propagated California Rainbow Trout Strains. California Fish and Game 66(1):17-24.
Descriptors: California; rainbow trout; fresh water fish; hatchery; artificial spawning; selective breeding
Genus Species: Salmo gairdneri

Buss, K., 1980. Photoperiod Control for Brood Trout. Aquaculture Magazine 6(2):45-48.
Descriptors: Photoperiodism; brood stock; problems; rainbow trout; fresh water fish; light; induced spawning; brook trout; anadromous fish; brown trout
Genus Species: Salmo gairdneri; Salvelinus fontinalis; Salmo trutta

Doroshov, D.I., 1977. The Present Status and Perspectives for Artificial Rearing and Acclimation of Sea and Brackish Water Fish in the USSR. Proceedings of the third Japan-Soviet Joint Symposium on Aquaculture, Nov. 1974, Tokyo, Japan, pp. 63-73.
Descriptors: USSR; brackish water; sea water; selective breeding; fresh water fish; Black Sea turbot; flatfish; marine fish; rainbow trout; striped bass; paddlefish; sea perch; ayu; sockeye salmon; anadromous fish; adaptability
Genus Species: Marone saxatilis; Polyodon spathula; Salmo gairdneri; Lateolabrax japonicus; Scophthalmus maeoticus; Plecoglossus altivelis; Oncorhynchus nerka

Eble, A.F., 1977. Integration of Thermal and Food Processing Residuals into a System for Commercial Culture of Freshwater Shrimp (Power Plant Waste Heat Utilization in Aquaculture). Vol. II. Public Service Electric and Gas Co. Research and Development Department. Final Report. NSF/RANN Grant No. S AEN 74-14079 AO1, GI-43925 2:1-165.
Descriptors: Giant freshwater prawn; waste water aquaculture; powerstation effluent; crustacean; stock density; pond culture; raceway culture; production; substrate; prawn; shrimp; intensive culture; rainbow trout; fresh water fish; nursery; size stocking; American eel; catadromous fish; weight; length; automatic device; temperature; oxygen; hatching; harvesting; season; flow; chlorine; thermal shock; light; growth; diet; dietary additive; pond; larva; management; disease; feeding; mortality; planning; food conversion; cytology; survival; New Jersey; protein; water

quality; juvenile; brood stock; water level; aquarium culture; food consumption; brine shrimp; heating; equipment; pellet; artificial food; stomach; fouling; food consumption; water column; artificial substrate; behavior; stress; mass mortality; parasite; disease detection; gill; liver; heart; intestinal tract; stomach; kidney; spleen; predation; cooling; water analysis; chemistry; heavy metal; bacteria; aquarium; labor
Genus Species: Macrobrachium rossenbergii; Salmo gairdneri

Eble, A.F., 1977. Power Plant Waste Heat Utilization in Aquaculture. First Annual Report. Appendix V. Status Report. (Performance Period: Nov.,1976—Nov., 1977) Public Service Electric and Gas Co. Research and Development Department. Status Report. NSF/RANN Grant No. S ENV 76-19854 AO1, PSE&G Grant RO-443 I-VIII, pp. 1-69.
Descriptors: Power station effluent; New Jersey; giant fresh water prawn; prawn; shrimp; crustacean; American eel; catadromous fish; rainbow trout; fresh water fish; stock density; production; growth; length; mortality; pellet; artificial food; raceway culture; intensiveculture; cannibalism; anadromous fish; management; planning; research; stocking; problems; well water; binder; dietary additive; live food; brine shrimp; pond culture; larva; juvenile; disease; disease treatment; weight; temperature; growth; aquarium culture; water quality; chemical; biochemical oxygen demand; laboratory; brood stock; feeder; equipment; closed system; striped bass
Genus Species: Macrobrachium rosenbergii; Anguilla rostrata; Salmo gairdneri; Morone saxatilis

Ebel, A.F., Stolpe, N.E., Evans, M.C., Deblois, N., Passanza, T., 1978. Power Plant Waste Heat Utilization in Aquaculture. Appendix VII. Status Report. I. Prawn Compartmentalization Experiment. II. Evaluation of Selected Systems of the "Proof of Concept" Facility. III. Renovation of Laboratory I, Prawn Broodstock Data and Experiments with Larval Macrobrachium. IV. Experiments with Early Juvenile Eels, Aquaculture Laboratories, Mercer Generating Station, Trenton, N.J. Public Service Electric and Gas Co. Research and Development Department. Status Report. NSF/RANN Grant No. ENV 76-19854 AO3, pp. 1-60.
Descriptors: Stocking; trout; fresh water fish; recirculated water; weight; harvesting; production; juvenile; American eel; catadromousfish; raceway culture; channel catfish; stock density; prawn; crustacean; aquarium culture; striped bass; anadromous fish; feeding; temperature; stress; power station effluent; oxygen; mortality; saprolegnia; fungus; infection; rainbow trout; giant freshwater prawn; New Jersey; water quality; management; chlorine; flow; brood stock; cost; food preparation; gas-bubble disease; disease treatment; disease control
Genus Species: Salmo gairdneri; Macrobrachium rosenbergii; Anguilla rostrata; Ictalurus punctatus; Morone saxatilis

Ehlinger, N.F., 1977. Selective Breeding of Trout for Resistance to Furunculosis. New York Fish and Game Journal 24(1):25-36.
Descriptors: Brook trout; anadromous fish; brown trout; fill disease; furunculosis; selective breeding; fresh water fish; resistance; survival; fry; artificial spawning; egg; sex; hatchery; brood stock; fungus; pond culture; symptom; aquarium culture
Genus Species: Salvelinus fontinalis; Salmo trutta

Erickson, J.D., 1981. American Trout Farming Marks 100 Years Plus—Still Growing. Aquaculture Magazine 7(3):14-17.
Descriptors: Fresh water fish; United States; rainbow trout; egg; brood stock; temperature; site selection; history; production; artificial food; pellet; Idaho; market; problems; fee fishing; Wisconsin; California; Colorado; Montana; Missouri; Washington; induced spawning; recreation
Genus Species: Salmo gairdneri

Gall, G.A.E., Gross, S.J., 1978. A Genetics Analysis of the Performance of Three Rainbow Trout Brood Stocks. Aquaculture 15(2):113-127.
Descriptors: Genetics; brood stock; rainbow trout; fresh water fish; reproduction; weight; fecundity; egg; stress; phenotype; size; California; production
Genus Species: Salmo gairdneri

Ghittino. P., 1977. Inspection and Certification of Fish for the International Control of Infectious Diseases of Salmonids. Proceedings from the International Symposium on Diseases of Cultured Salmonids. Seattle, Washington, April 4-6, 1977. Sponsored by Tavolek, Inc., pp. 74-86.
Descriptors: Rainbow trout; fresh water fish; egg; economics; selective breeding; hatchery; raceway culture; nutrition; artificial food; pellet; market; redmouth disease; Europe; mortality; whirling disease; furunculosis; vibrio; disease control; virus; brown trout; gill disease; fry; inspection; legal aspect; management
Genus Species: Salmo gairdneri; Salmo trutta

Gjedrem, T., Naevdal, G., 1979. Research on Quantitative Genetics on Salmonids in Norway. International Council for the Exploration of the Sea, Mariculture Committee Paper F:22, 8 pp.
Descriptors: Research; genetics; Norway; economics; selective breeding; egg; Atlantic salmon; rainbow trout; anadromous fish; fresh water fish
Genus Species: Salmo gairdneri; Salmo salar

Gunnes, K., Gjedrem, T., 1978. Selection Experiments with Salmon IV. Growth of Atlantic Salmon During Two Years in the Sea. Aquaculture 15(1):19-33.
Descriptors: Atlantic salmon; anadromous fish; brood stock; weight; length; genetics; growth; phenotype; cage culture; feeding; management
Genus Species: Salmo salar

Jamieson, A., 1980. The Cage Rearing of Rainbow Trout in a Brackish Water Pond in Newfoundland, 1978. Canadian Industry Report of Fisheries and Aquatic Sciences 115:i-iv, 1-18.
Descriptors: Cage culture; rainbow trout; fresh water fish; pondculture; brackish water; pilot program; Canada; Newfoundland; brood stock; feasibility study; growth; weight; waste; construction detail; cage; equipment; raceway culture; pellet; artificial food; pond; feed composition; food conversion; survival; temperature; stock density; oxygen
Genus Species: Salmo gairdneri

Kincaid, H.L., 1979. Development of Standard Reference Lines of Rainbow Trout. Transactions of the American Fisheries Society 108(5):457-461.
Descriptors: Rainbow trout; fresh water fish; genetics; research;Wyoming; egg; hatching; fry; survival; growth; heritability; selective breeding; weight
Genus Species: Salmo gairdneri

Kincaid, H.L., Bridges, W.R., Von Limbach, B., 1977. Three Generations of Selection for Growth Rate in Fall-Spawning Rainbow Trout. Transactions of the American Fisheries Society 106(6):621-628.
Descriptors: Selective breeding; growth; rainbow trout; freshwaterfish; genetics; Wyoming; weight; hatching; fry; survival; aquarium culture; phenotype
Genus Species: Salmo gairdneri

Kinunen, W., Moring, J.R., 1978. Origin and Use of Oregon Rainbow Trout Brood Stocks. The Progressive Fish-Culturist 40(3):87-89.
Descriptors: Brood stock; Oregon; rainbow trout; fresh water fish; hatchery; genetics; production
Genus Species: Salmo gairdneri

Klopfenstein, D., Klopfenstein, I., 1981. Farm Ponds Flourish on Montana Prairie. Fish Farming International 8(2):18-19.
Descriptors: Montana; pond culture; brood stock; rainbow trout; fresh water fish; fresh water shrimp; crustacean
Genus Species: Salmo gairdneri

Klupp, R., 1979. Genetic Variance for Growth in Rainbow Trout (*Salmo gairdneri*). Aquaculture 18(2):123-134.
Descriptors: Genetics; growth; rainbow trout; fresh water fish; selective breeding; hatchery; trough culture; weight
Genus Species: Salmo gairdneri

Loginova, G.A., Krasnoperova, S.V., 1982. An Attempt at Crossbreeding Atlantic Salmon and Pink Salmon (Preliminary Report). Aquaculture 27(4):329-337.
Descriptors: Atlantic salmon; pink salmon; anadromous fish; hybrid; selective breeding; chromosome; embryo; development; abnormality
Genus Species: Salmo salar; Oncorhynchus gorbuscha

Lynch, T., 1980. Abandoned Ozark Village Developed into Unique Trout Fee Fishing Facility. Aquaculture Magazine 6(3):28-31.
Descriptors: Missouri; fee fishing; commercial firm; rainbow trout; fresh water fish; hatchery; raceway culture; feeding; brood stock; artificial fertilization; hatching; well water; water supply; fry; pond culture; recreation
Genus Species: Salmo gairdneri

MacGregor, R.B., MacCrimmon, H.R., 1977. Evidence of Genetic and Environmental Influences on Meristic Variation in the Rainbow Trout, *Salmo gairdneri* Richardson. Environmental Biology of Fishes 2(1):25-33. Dr. W. Junk, B.V. Publisher, The Hague, The Netherlands.
Descriptors: Rainbow trout; fresh water fish; temperature; morphology; brood stock; selective breeding; artificial spawning; incubator; aquarium culture; artificial food; pellet; anatomy; genetics; heritability
Genus species: Salmo gairdneri

McCarthy, D.H., 1977. Present Status of Aeromonas Infections. Proceedings from the International Symposium on Diseases of Cultured Salmonids. Seattle, Washington, April 4-6, 1977. pp. 182-189. Sponsored by Tavolek, Inc.
Descriptors: Aeromonas; bacteria; disease; pond culture; infection; transmission; survival; furunculosis; infection; disease treatment; tetracycline; brown trout; contamination; rainbow trout; fresh water fish; water quality; coho salmon; anadromous fish; feeding; kidney; brood stock; streptomycin; identification; species isolation
Genus Species: Salmo gairdneri; Oncorhynchus kisutch; aeromonas salmonicida

McIntyre, J.D., 1977. Heritable Tolerance of Disease in Salmonids. Proceedings from the International Symposium on Diseases of Cultured Salmonids. Seattle, Washington, April 4-6, 1977. pp. 87-90. Sponsored by Tavolek, Inc.
Descriptors: Tolerance; disease; selective breeding; hatchery; genetics; heritability; vibrio; Atlantic salmon; anadromous fish; chinook salmon; sockeye salmon; virus; necrosis; brook trout; fresh water fish; brown trout; adaptability
Genus Species: Salmo salar; Onchorhynchus tshawytscha; Oncorhynchus nerka; Salvelinus fontinalis; Salmo trutta

Mighell, J.L., 1981. Culture of Atlantic Salmon, *Salmo salar*, in Puget Sound. Marine Fisheries Review 43(2):1-8.
Descriptors: Atlantic salmon; anadromous fish; Puget Sound; restoration; brood stock; fry; smolt; weight; size; spawning; survival; cage culture; pilot program; growth; resistance; incubation; incubator; equipment; substrate; temperature; salinity; sea water; fresh water; adaptability; disease treatment; maturity
Genus Species: Salmo salar

Naevdal, G., Leroy, R., Moller, D., 1979. Variation in Growth Rate and Age at First Maturation in Rainbow Trout. International Council for the Exploration of the Sea, Mariculture Committee Paper, F:21, 13 pp.
Descriptors: Growth; age; maturity; rainbow trout; fresh water fish; selective breeding; genetics; length; weight; heritability; hatchery; Norway; cage culture
Genus Species: Salmo gairdneri

Rao, B.S., Chandrasekaran, G., 1978. Preliminary Report on Hybridization Experiments in Trout Growth and Survival of F_1 Hybrids. Aquaculture 15(3):297-300.
Descriptors: Hybrid; growth; survival; fresh water fish; rainbow trout; genetics; egg; fry; selective breeding; development; gonad; hatchery; India
Genus Species: Salmo gairdneri

Refstie, T., Steine, T.A., Gjedrem, T., 1977. Selection Experiments with Salmon. II. Proportion of Atlantic Salmon Smoltifying at One Year of Age. Aquaculture 10(3):231-242.
Descriptors: Atlantic salmon; anadromous fish; smolt; brood stock; weight; heritability; genetics
Genus Species: Salmo salar

Refstie, T., Steine, T.A., 1978. Selection Experiments with Salmon. III. Genetic and Environmental Sources of Variation in Length and Weight of Atlantic Salmon in the Freshwater Phase. Aquaculture 14(3):221-234.
Descriptors: Atlantic salmon; anadromous fish; length; weight; ecology; genetics; fresh water; Norway; brood stock; aquarium culture; growth; model; phenotype; genotype
Genus Species: Salmo salar

Ritter, J.A., Carey, T.G., 1980. Salmon Ranching. Chapter 7. Salmon Ranching in the Atlantic Maritime Provinces of Canada. Salmon Ranching, pp. 109-130. Academic Press.
Descriptors: Canada; anadromous fish; stocking; hatchery; spawning; incubation; hatching; pond culture; fry; selective breedng; open water culture; planning; Atlantic salmon
Genus Species: Salmo salar

Ryman, N., Stahl, G., 1980. Genetic Changes in Hatchery Stocks of Brown Trout (*Salmo trutta*). Canadian Journal of Fisheries and Aquatic Sciences 37(1):82-87.
Descriptors: Genetics; hatchery; brown trout; fresh water fish; Sweden; electrophoresis; stocking; inbreeding; brood stock
Genus Species: Salmo trutta

Sedgwick, S.D., 1978. Trout Farming Handbook. 7. Brood Stock. Trout Farming Handbook, pp. 85-94. Scholium International, Inc.
Descriptors: Rainbow trout; fresh water fish; brood stock; spawning; season; resistance; growth; temperature; maturity; inbreeding; egg; disease control
Genus Species: Salmo gairdneri

Sedgwick, S.D., 1978. Trout Farming Handbook. 14. Cages for Freshwater Trout Production. Trout Farming Handbook, pp. 158-164. Scholium International, Inc.
Descriptors: Rainbow trout; fresh water fish; cage culture; siteselection; equipment; net; construction detail; feeding; grading; brood stock
Genus Species: Salmo gairdneri

Smith, C.E., Osborne, M.D., Piper, R.G., Dwyer, W.P., 1979. Effect of Diet Composition on Performance of Rainbow Trout Brood Stock During a Three-Year Period. The Progressive Fish Culturist 41(4):185-188.
Descriptors: Diet; feed composition; rainbow trout; fresh water fish; brood stock; energy; protein; fecundity; age; mortality; vitamin; weight; cost
Genus Species: Salmo gairdneri

Sutterlin, A.M., Harman, P., Young, B., 1978. Precocious Sexual Maturation in Atlantic Salmon (*Salmo salar*) Postsmolts Reared in a Seawater Impoundment. Journal of the Fisheries Research Board of Canada 35(9):1269-1272.
Descriptors: Maturity; development; Atlantic salmon; anadromous fish;smolt; sea water; growth; salinity; tolerance; cage culture; brood stock
Genus Species: Salmo salar

Thorpe, J.E., Morgan, R.I.G., 1978. Parental Influence on Growth Rate, Smolting Rate and Survival in Hatchery Reared Juvenile Atlantic Salmon, *Salmo salar*. Journal of Fish Biology 13(5):549-556.
Descriptors: Growth; survival; hatchery; juvenile; Atlantic salmon;anadromous fish; smolt; length; mortality; genetics; heritability; brood stock
Genus Species: Salmo salar

Walsh, D. (Chairman), 1978. Aquaculture in the United States. Constraints and Opportunities. Appendix B. Current Status of Genetics and Selective Breeding in Major Aquaculture Species. National Academy of Sciences, pp. 103-107.
Descriptors: Genetics; selective breeding; carp; fresh water fish; lake trout; brook trout; rainbow trout; channel catfish; Tilapia; giant freshwater prawn; shrimp; crayfish; crustacean; American lobster; mollusk; eastern oyster; northern quahog; bay scallop; seaweed
Genus Species: Cyprinus carpio; Salvelinus namaycush; Salvelinusfontinalis; Salmo gairdneri; Ictalurus punctatus; Tilapia; Macrobrachium rosenbergii; Homarus americanus; Mercenaria; Argopecten irradians; Crassostrea virginica

Salmonid Genetic Studies

Allen, S.K., Jr., Stanley, J.G., 1979. Polyploid Mosaics Induced by Cytochalasin B in Landlocked Atlantic Salmon, *Salmo salar*. Transactions of the American Fisheries Society 108(5):462-466.
Descriptors: Atlantic salmon; anadromous fish; genetics; reproduction; population control; fecundity; Maine; chromosome
Genus Species: Salmo salar

Allen, S.K., Jr., Stanley, J.G., 1981. Mosaic Polyploid Atlantic Salmon (*Salmo salar*) Induced by Cytochalasin B. The Early Life History of Fish: Recent Studies, the Second ICES Symposium, Woods Hole, April 2-5, 1979, Conseil International pour L'Exploration de la Mer, Rapports et Proces-Verbaux des Reunions 178, 509-510.
Descriptors: Atlantic salmon; anadromous fish; genetics; dosage; meiosis; chemical; chromosome; sterility; incubation
Genus Species: Salmo salar

Andrews, J.W., 1979. A Summary of Aquaculture Research at the Skidaway Institute and the University of Georgia's Coastal Plain Experiment Station. The University of Georgia, College of Agriculture, Experiment Station, Research Bulletin 247:1-98.
Descriptors: Research; channel catfish; fresh water fish; intensive culture; diet; cost; food conversion; genetics; site selection; raceway culture; shrimp; crustacean; artificial food; problems; feeding; larva; brine shrimp; mixed culture; pond culture; effluent; rainbow trout; striped bass; American shad; anadromous fish; goldfish; baitfish; minnow; tilapia; stock density; pH; size; temperature; light; photoperiodism; oxygen; food consumption; Georgia; nutrition; lipid; protein; methionine; amino acid; dietary additive; arginine; pellet; fry; vitamin; mineral; survival; giant freshwater prawn
Genus Species: Ictalurus punctatus; Salmo gairdneri; Morone saxatilis; Alosa sapidissima; Carassius auratus; Tilapia; Macrobrachium rosenbergii

Avault, J.W., Jr., 1983. Maximizing Production and Profit; Efficient Studies. Aquaculture Magazine 9(5):42-44.
Descriptors: Production; profit; species selection; mollusk; crustacean; red swamp crayfish; tilapia; fresh water fish; growth; channel catfish; bluegill; pond culture; far east; Europe; Israel; rainbow trout; United States; genetics; largemouth bass; black bass; carp; intensive culture; hybrid; cage culture; aquarium culture; blue catfish
Genus Species: Tilapia melanopleura; Tilapia mossambica; Ictalurus punctatus; Ictalurus furcatus; Lepomis macrochirus; Salmo gairdneri; Micropterus salmoides; Cyprinus carpio; Hypophthalmichthys molitrix; Hypophthalmichthys nobilis; Ctenopharyngodon idellus

Ayles, G.B., Bernard, D., Hendzel, M., 1979. Genetic Differencies in Lipid and Dry Matter Content Between Strains of Rainbow Trout (*Salmo gairdneri*) and Their Hybrids. Aquaculture 18(3):253-262.
Descriptors: Genetics; lipid; rainbow trout; fresh water fish; hybrid; phenotype; Canada; size; growth; protein; pond culture
Genus Species: Salmo salar

Bry, C., 1981. Temporal Aspects of Macroscopic Changes in Rainbow Trout (*Salmo gairdneri*) Oocytes Before Ovulation and of Ova Fertility during the Post-Ovulation Period; Effect of Treatment with 17α-Hydroxy-20β-Dihydroprogesterone. Aquaculture 24(1,2):153-160.
Descriptors: Rainbow trout; fresh water fish; ovulation; oocyte; ovum; fecundity; hormone; injection; induced spawning; viability; steroid
Genus Species: Salmo gairdneri

Chevassus, B., 1979. Hybridization in Salmonids: Results and Perspectives. Aquaculture 17(2):113-128.
Descriptors: Hybrid; research; brook trout; lake trout; arctic char; Dolly Varden trout; anadromous fish; fresh water fish; fecundity; selective breeding
Genus Species: Salvelinus fontinalis; Salvelinus namaycush; Salvelinus alpinus; Salvelinus malma

Fraser, J.M., 1981. Comparative Survival and Growth of Planted Wild, Hybrid, and Domestic Strains of Brook Trout (*Salvelinus fontinalis*) in Ontario Lakes. Canadian Journal of Fisheries and Aquatic Sciences 28(12):1672-1684.
Descriptors: Survival; growth; hybrid; wild stock; brook trout; anadromous fish; recapture; Canada; lake; water; hatchery; condition; selective breeding; transplanting
Genus Species: Salvelinus fontinalis

Glover, M., Low, C., 1982. Salmonid Enhancement Program. 1981 Annual Report Summary. Salmonid Enhancement Program, Department of Fisheries and Oceans, Province of British Columbia, 22 pp.
Descriptors: Canada; British Columbia; restoration; resources; anadromous fish; fishing; labor; education; production; economics; ecology; sockeye salmon; chum salmon; pink salmon; coho salmon; chinook salmon; cutthroat trout; rainbow trout; fresh water fish; hatchery; open water culture; research; planning; incubation
Genus Species: Salmo gairdneri; Salmo clarki; Oncorhynchus nerka; Oncorhynchus keta; Oncorhynchus gorbuscha; Oncorhynchus kisutch; Oncorhynchus tsawytscha

Gold, J.R., Pipkin, R.E.; Gall, G.A.E., 1979. Notes on a Hybridization Experiment Between Rainbow and Golden Trout. California Fish and Game 65(3):179-183.
Descriptors: Hybrid; rainbow trout; fresh water fish; golden trout; genetics; morphology; mortality
Genus Species: Salmo gairdneri; salmo aguabonita

Gunnes, K., Gjedrem, T., 1981. A Genetic Analysis of Body Weight and Length in Rainbow Trout Reared in Seawater for 18 Months. Aquaculture 24(1,2):161-174.
Descriptors: Genetics; weight; length; rainbow trout; fresh water fish; sea water; Norway; cage culture; phenotype; heritability; growth
Genus Species: Salmo gairdneri

Hancock, R. (Editor), 1982. Disease-Free Eggs. World Fishing 31(7):16.
Descriptors: Egg; United Kingdom; rainbow trout; fresh water fish; disease control; genetics; selective breeding; hatchery; induced spawning; ultraviolet; biofilter; equipment; filter; brood stock; commercial firm
Genus Species: Salmo gairdneri

Harache, Y., Boeuf, G., 1979. Reproduction in Captivity of Coho Salmon (*Oncorhynchus kisutch*). International Council for the Exploration of the Sea, Mariculture Committee Paper F:51, 11 pp.
Descriptors: Reproduction; coho salmon; anadromous fish; aquarium culture; sea water; fresh water; growth; mortality; fecundity; ovulation; brood stock; mortality; gonad; development; incubation; erythromycin; antibiotic; drug
Genus Species: Oncorhynchus kisutch

Hershberger, W.K., 1979. The Relationship of Genetics Work on the Anadromous Salmonids of the North Pacific to Culture Problems. International Council for the Exploration of the Sea, Mariculture Committee Paper F:33, 1-12.
Descriptors: Genetics; anadromous fish; problems; rainbow trout; fresh water fish; management; reproduction; cage culture; open water culture; coho salmon; pink salmon; chum salmon; sockeye salmon; chinook salmon
Genus Species: Salmo gairdneri; Oncorhynchus gorbuscha; Oncorhynchus keta; Oncorhynchus kisutch; Oncorhynchus nerka; Oncorhynchus tsawytscha

Hershberger, W.K.; Iwamoto, R.N., Saxton, A.M., 1982. Genetic Potential for Fresh- and Seawater Growth of Net-Pen Cultured Coho Salmon. Proceedings of the North Pacific Aquaculture Symposium, Anchorage, Alaska, August 18-21, 1980; Newport, Oregon, August 25-27, 1980. Alaska Sea Grant Report 82-2, pp. 185-192.
Descriptors: Genetics; fresh water; sea water; growth; cage culture; coho salmon; anadromous fish; brood stock; selective breeding; heritability
Genus Species: Oncorhynchus kisutch

Hjul, P. (Editor), 1982. Sea Cage Tests in Cyprus. Fish Farming International 9(1):7.
Descriptors: Cage culture; Cyprus; gilt head bream; sea bass; rabbitfish; artificial food; induced spawning; diet; food conversion; marine fish; shrimp; crustacean; cutthroat trout; fresh water fish; growth
Genus Species: Sparus auratus; Dicentrarchus labrax; Siganus rivulatus; Penaeus kerathurus; Penaeus japonicus; Salmo clarki

Johnstone, R., Simpson, T.H.; Youngson, A.F., Whitehead, C., 1979. Sex reversal in Salmonid Culture. Part II. The Progeny of Sex-Reversed Rainbow Trout. Aquaculture 18(1):13-19.
Descriptors: Sex reversal; rainbow trout; hormone; reproduction; testis; mortality; sex; fresh water fish
Genus Species: Salmo gairdneri

Kinghorn, B.P., 1983. A Review of Quantitative Genetics in Fish Breeding. Aquaculture 31(2,3,4):283-304.
Descriptors: Genetics; species selection; selective breeding; inbreeding; abnormality; genotype; ecology; heritability; rainbow trout; channel catfish; tilapia; brown trout; fresh water fish; Atlantic salmon arctic char; sea trout; anadromous fish; food conversion; survival; growth; production; weight; resistance; tolerance; hybrid; biological competition; cryogenic preservation; cytology
Genus Species: Salmo gairdneri; Ictalurus punctatus; Tilapia nilotica; Salmo trutta; Salmo salar; Salvelinus alpinus

Lemoine, H.L., Jr., Smith, L.T., 1980. Polyploidy Induced in Brook Trout by Cold Shock. Transactions of the American Fisheries Society 109(6):626-631.
Descriptors: Brook trout; anadromous fish; thermal shock; chromosome; genetics; mortality; induced spawning; egg
Genus Species: Salvelinus fontinalis

Lill, A.F., Eng, P., 1981. A Perspective on the Salmonid Enhancement Program in British Columbia. Proceedings of the Bio-Engineering Symposium for Fish Culture, Traverse City, Michigan, Oct. 16-18, 1979, FCS of AFS, pp. 274-281.
Descriptors: British Columbia; government agency; bioengineering; salmon; anadromous fish; incubation; hatchery; spawning; fresh water; ecology; management
Genus Species: Oncorhynchus tshawytscha; Oncorhynchus keta; Oncorhynchus gorbuscha; Oncorhynchus nerka

MacKay, K.T., Van Toever, W., 1981. An Ecological Approach to a Water Recirculating System for Salmonids: Preliminary Experiences. Proceedings of the Bio-Engineering Symposium for Fish Culture, Traverse City, Michigan, Oct. 16-18, 1979, FCS of AFS, pp. 249-258.
Descriptors: Ecology; recirculated water; algae; solar aquaculture; aeration; rainbow trout; fresh water fish; water quality; growth; survival; oxygen; mortality; design; energy; temperature; pH; alkalinity; nitrogen; Canada
Genus Species: Salmo gairdneri

Nic, J., Ingols, R., 1981. Cause of Trout Mortality in Hatcheries. The Progressive Fish-Culturist 43(1):32-36.
Descriptors: Mortality; hatchery; fresh water fish; manganese; metal; rainbow trout; ecology; Arkansas; Georgia; water quality; oxygen; problems
Genus Species: Salmo gairdneri

Okada, H., Matumoto, H., Yamazaki, F., 1979. Functional Masculinization of Genetic Females in Rainbow Trout. Bulletin of the Japanese Society of Scientific Fisheries 45(4):413-419.
Descriptors: Rainbow trout; fresh water fish; sex reversal; hormone; mortality; sex ratio; genetics
Genus Species: Salmo gairdneri

Onozato, H., 1982. The "Hertwig Effect" and Gynogenesis in Chum Salmon, *Oncorhynchus keta*, Eggs Fertilized with 60°C Gamma-Ray Irradiated Milt. Bulletin of the Japanese Society of Scientific Fisheries 48(9):1237-1244 (In Japanese with English abstract and tables).
Descriptors: Dosage, gynogenesis; gamma ray; chum salmon; anadromous fish; egg; irradiation; fertilization; semen; sperm; cobalt; metal; survival; embryo; chromosome; genetics; development
Genus Species: Oncorhynchus keta

Refstie, T., 1980. Genetic and Environmental Sources of Variation in Body Weight and Length of Rainbow Trout Fingerlings. Aquaculture 19(4):351-357.
Descriptors: Genetics; weight; length; fry; rainbow trout; fresh water fish; phenotype; heritability; aquarium culture; stock density
Genus Species: Salmo gairdneri

Reinitz, G.L., Orme, L.E., Hitzel, F.N., 1979. Variations of Body Composition and Growth Among Strains of Rainbow Trout. Transactions of the American Fisheries Society 108(2):204-207.
Descriptors: Body composition; growth; rainbow trout; fresh water fish; food conversion; genetics; genotype; diet; ash; protein
Genus Species: Salmo gairdneri

Schreck, C.B., Fowler, L.G., 1982. Growth and Reproductive Development in Fall Chinook Salmon: Effects of Sex Hormones and Their Antagonists. Aquaculture 26(3,4):253-263.
Descriptors: Growth; reproduction; chinook salmon; anadromous fish; hormone; juvenile; steroid; development; gonad; aquarium culture; food conversion; length
Genus Species: Oncorhynchus tshawytscha

Smith, L.T., Lemoine, H.L., 1979. Colchicine-Induced Polyploidy in Brook Trout. The Progressive Fish-Culturist 41(2):86-88.
Descriptors: Brook trout; anadromous fish; genetics; colchicine; embryo; mortality; chromosome; reproduction
Genus Species: Salvelinus fontinalis

Stanley, J.G., 1981. Manipulation of Developmental Events to Produce Monosex and Sterile Fish. The Early Life History of Fish: Recent Studies, The Second ICES Symposium, Woods Hole, April 2-5, 1979. Conseil International pour L'Exploration de la Mer, Rapports et Proces-Verbaux des Reunions 178:485-491.
Descriptors: Genetics; grass carp; fresh water fish; gynogenesis; monosex culture; thermal shock; Atlantic salmon; anadromous fish; brook trout; sex reversal; irradiation
Genus Species: Ctenopharyngodon idellus; Salmo salar; Salvelinus fontinalis

Stuart-Kregor, P.A.C., Sumpter, J.P., Dodd, J.M., 1981. The Involvement of Gonadotrophin and Sex Steroids in the Control of Reproduction in Parr and Adults of Atlantic Salmon, *Salmo salar l.* Journal of Fish Biology 18(1):59-72.
Descriptors: Gonadotropin; steroid; reproduction; Atlantic salmon; anadromous fish; hormone; androgen; pituitary; United Kingdom; maturity; gonad
Genus Species: Salmo salar

Sutterlin, A.M., 1983. A Review of Technical Data at Hopeall Trout Farm. Unpublished report, Marine Sciences Research Laboratory, 36 pp.
Descriptors: Hatchery; temperature; pond; growth; brood stock; maturity; fecundity; size; disease; season; mortality; moist diet; artificial food; salinity; rainbow trout; fresh water fish; brackish water; oxygen; site selection; egg; fungus; virus; parasite; bacteria; vibrio; pseudomonad; aeromonas; Canada; Newfoundland
Genus Species: Salmo gairdneri

Tacon, A.G.J., Desilva, S.S., 1983. Mineral Composition of Some Commercial Fish Feeds Available in Europe. Aquaculture 31(1):11-20.
Descriptors: Mineral; feed composition; Europe; rainbow trout; fresh water fish; Atlantic salmon; anadromous fish; European eel; catadromous fish; diet; moisture; pH; ash; sodium chloride; calcium; phosphorus; potassium; sodium; magnesium; trace element; fry; larva; brood stock
Genus Species: Salmo gairdneri; Salmo salar; Anguilla anguilla

Thorgaard, G.H., Jazwin, M.E., 1981. Polyploidy Induced by Heat Shock in Rainbow Trout. Transactions of the American Fisheries Society 110(4):546-550.
Descriptors: Rainbow trout; fresh water fish; genetics; chromosome; egg; thermal shock; temperature; survival; Washington
Genus Species: Salmo gairdneri

Wedemeyer, G.A., Saunders, R.L., Clarke, W.C., 1980. Environmental Factors Effecting Smoltification and Early Marine Survival of Anadromous Salmonids. Marine Fisheries Review 42(6):1-14.
Descriptors: Ecology, survival; anadromous fish; open water culture; Atlantic salmon; restoration; smolt; rainbow trout; fresh water fish; coho salmon; salinity; behavior; tolerance; growth; condition; body composition; migration; temperature; photoperiodism; disease treatment; hatchery; mortality; age; size; stress; kidney disease
Genus Species: Salmo salar; Salmo gairdneri; Oncorhynchus kisutch

Wray, T., 1979. Shearwater: British Big Company Success in Fish Farming. Fish Farming International 6(3):24-28.
Descriptors: United Kingdom; commercial firm; rainbow trout; feeding; fresh water fish; processing; salinity; size; intensive culture; species selection; turbot; flatfish; marine fish; live food; rotifer; zooplankton; aquarium culture; market; induced spawning; artificial food; moist diet
Genus Species: Salmo gairdneri; Scophthalmus maximus

Zaugg, W.W., 1982. Some Changes in Smoltification and Seawater Adaptability of Salmonids Resulting from Environmental and Other Factors. Aquaculture 28(1,2):143-151.
Descriptors: Sea water; adaptability; ecology; hatchery; Washington; chinook salmon; migration; coho salmon; anadromous fish; blood; hematology; smolt; mortality
Genus Species: Oncorhynchus kisutch; Oncorhynchus tshawytscha

Zohar, Y., Billard, R., 1979. New Data on the Possibilities of Controlling Reproduction in Teleost Fish by Hormonal Treatment. CNEXO. Actes de Colloques 8:111-123.
Descriptors: Reproduction; hormone; rainbow trout; fresh water fish; induced spawning; ovulation; northern pike; carp; gilt head bream; gonadotropin
Genus Species: Salmo gairdneri; Esox lucius; Cyprinus carpio; Sparus auratus

Symposium Participants

Katsumi Aida
Dept. of Fisheries
Faculty of Agriculture
University of Tokyo
1-1-1 Yayoi, Bunkyo-ku
Tokyo, Japan

R. Alderson
Marine Harvest Ltd.
Lochailort, Inverness-shire
Scotland PH38 4LZ
United Kingdom

Marianna Alexandersdottir
Fisheries Research Institute, WH-10
University of Washington
Seattle, WA 98195

Brian J. Allee
Prince William Sound Regional
Aquaculture Corp.
Cordova, AK 99574

Standish K. Allen
School of Fisheries, WH-10
University of Washington
Seattle, WA 98195

Donald F. Amend
S.S.R.A.A.
Box 6916
Ketchikan, AK 99901

Aven M. Andersen
Fishery Management Operations Division
National Marine Fisheries Service
3300 Whitehaven St. NW
Washington, DC 20235

Ron J. Anderson
Bioproducts, Inc.
Box 429
Warrenton, OR 97146

Bruce Bachen
N.S.R.A.A.
P.O. Box 2606
Sitka, AK 99835

Randy Baker
Freshwater Institute
501 University Crescent
Winnipeg, Manitoba R3T 2N6
Canada

R. Martin Ball
2060 - 23rd E.
Seattle, WA 98112

Robert Bastian
U.S. E.P.A.
Office of Water (WH-556)
401 M St. SW
Washington, DC 20460

Roy E. Beaty
708 SW 13th
Corvallis, OR 97333

C. Dale Becker
Battelle Northwest
P.O. Box 999
Richland, WA 99352

Tillmann Benfey
Marine Sciences Research Laboratory
Memorial University
St. John's, Newfoundland A1C 5S7, Canada

Cliff Bengston
Tulalip Tribes
10610 Waterworks Rd.
Tulalip, WA 98270

Bill Berestoff
Kodiak Aqauculture Assn.
Box 2194
Kodiak, AK 99615

Jim Bergeron
Clatsop County Extension
P.O. Box 207
Astoria, OR 97103

Sam Bertoni
Alaska Dept. Fish & Game
Rm. 301, 230 S. Franklin St.
Juneau, AK 99801

Evelyn Biggs-Chisum
Alaska Fisherman's Journal
1115 NW 46th
Seattle, WA 98107

Rolland Billard
Laboratoire de Physiologie
des Poissons
Institut National de la
Recherche Agronomique
78350 Jouy-en-Josas, France

Gary Birch
8240 St. Alban's Rd.
Richmond, B.C. V6Y 2K9
Canada

Roy T. Blaz
New Mexico Dept. of
Fish & Game
Parkview Hatchery
Los Ojos, NM 87551

John P. Blum
Makah Tribal Council
P.O. Box 115
Neah Bay, WA 98357

Ed Brauer
DomSea
4398 W. Old Belfair Hwy.
Bremerton, WA 98312

Scott Brewer
School of Fisheries, WH-10
University of Washington
Seattle, WA 98195

Niall Bromage
Fish Culture Unit
University of Aston
Birmingham B4 7ET
United Kingdom

Dan Brown
Mt. Lassen Trout Farm
Rt. 5, Box 36
Red Bluff, CA 96080

Robert A. Busch
Clear Springs Trout Co.
P.O. Box 712
Buhl, ID 83316

Russell W. Cahill
Washington Dept. of Fisheries
115 General Administration Bldg.
Olympia, WA 98506

Gary W. Camenisch
Mo. Dept. of Conservation
666 Primrose Ave.
Springfield, MO 65807

Jim Cardwell
#302 - 2555 Discovery St.
Vancouver, B.C., Canada

Leon Carl
Alberta Fish & Wildlife Division
North Tower, Petroleum Plaza
9945 - 108th St.
Edmonton, Alberta T5K 2G6
Canada

Bob Carson
NE 1024 'D' St., #2
Pullman, WA 99163

Antonio Carlos Carvalho Filho
Dept. Fisheries & Oceans
West Vancouver Laboratory
4160 Marine Drive
West Vancouver, B.C.
V7V 1N6 Canada

Hal Caswell
Biology Dept.
Woods Hole Oceanographic Institute
Woods Hole, MA 02543

Jeff Cederholm
Dept. of Natural Resources
RR 1, Box 1375
Forks, WA 98331

Huei-Mei Chen
Food Science Dept.
Louisiana State University
Baton Rouge, LA 70803

Kenneth K. Chew
School of Fisheries, WH-10
University of Washington
Seattle, WA 98195

Craig Clarke
Fisheries & Oceans
Pacific Biological Station
Nanaimo, B.C. V9R 5K6 Canada

Joe Cloud
Dept. of An. Sci.
495 AnSci/Vet. Medicine
University of Minnesota
1988 Fitch Avenue
St. Paul, MN 55108

Fred S. Conte
659 Fillmore
Davis, CA 95616

Faye Cowie
302 Charlotte St., Apt. 2
Fredericton, New Brunswick
E3B 1L7 Canada

Tom Craig
Physics Department
University of Guelph
Guelph, Ontario
N1G 2W1 Canada

Jean Cramer
336 NW 48th
Seattle, WA 98107

Laurence Crim
Marine Sciences Research Laboratory
Memorial University
St. John's, Newfoundland
A1C 5S7 Canada

Anthony W. Crowe
New Zealand Salmon Co., Ltd.
P.O. Box 8346
Riccarton, New Zealand

Steve Damm
10432 - 61st S.
Seattle, WA 98178

Brian Gilbert D'Aoust
Common Sensing, Inc.
7595 Finch Rd. NE
Bainbridge Is., WA 98110

Bill Davidson
Sheldon Jackson
P.O. Box 479
Sitka, AK 99835

Jac Davies
7325 - 14th NW
Seattle, WA 98107

Robert Davis
Alaska Fish & Game
333 Raspberry Rd.
Anchorage, AK 99502

William R. Davis
Washington Sea Grant
University of Washington
3716 Brooklyn Ave. NE
Seattle, WA 98105

Peter G. Daye
Daye Atlantic Salmon Corporation
P.O. Box 1081
Armdale, Nova Scotia
B3L 4L5 Canada

Michael B. Dell
Grant County P.U.D.
30 'C' St. SW
P.O. Box 878
Ephrata, WA 98823

Ed Devlin
Pathology, D-405 Health Sciences
University of Washington, SM-30
Seattle, WA 98195

Nancy Diamond
% Dr. David Hankin
Dept. of Fisheries
Humboldt State University
Arcata, CA 95521

Walton W. Dickhoff
Dept. of Zoology, NJ-15
University of Washington
Seattle, WA 98195

Jeff Dickison
School of Fisheries, WH-10
University of Washington
Seattle, WA 98195

Edward M. Donaldson
Fisheries Research Branch
4160 Marine Drive
West Vancouver, B.C.
V7V 1N6 Canada

James Donaldson
Coast Oyster Co.
P.O. Box 327
Quilcene, WA 98376

Lauren Donaldson
School of Fisheries, WH-10
University of Washington
Seattle, WA 98195

Paul Dorn
Suquamish Fisheries
P.O. Box 498
Suquamish, WA 98392

Carol Dorworth
308 Fernow Hall
Cornell University
Ithaca, NY 14853

Thurston Dotson
Dept. of Fish & Wildlife
Box 508
Big Timber, MT 59011

Donald Drake
Moore-Clark Co., Inc.
P.O. Box M
LaConner, WA 98257-0378

John A. Driver
Marquette State Fish Hatchery
488 Cherry Creek
Marquette, MI 49855

Ben Drucker
National Marine Fisheries Service
Office of Research Investigations
Washington, DC 20235

Gary Duker
Fisheries Research Institute, WH-10
University of Washington
Seattle, WA 98195

Henry F. Eaton
Kodiak Aquaculture Assn.
Box 1423
Kodiak, AK 99615

Louie S. Echols
Washington Sea Grant
University of Washington
3716 Brooklyn Ave. NE
Seattle, WA 98105

Hillary Egna
611 NW Van Buren, Apt. 4
Corvallis, OR 97330

Al Erdahl
University of Minnesota
495 AnSc/Vet Med
1988 Fitch Ave.
St. Paul, MN 55108

Mike Erho
Douglas County P.U.D.
1151 Valley Mall Parkway
E. Wenatchee, WA 98801

J. David Erickson
Clear Springs Trout Co.
Box 712
Buhl, ID 83316

Randy Eshenroder
Great Lakes Fishery Commission
1451 Green Road
Ann Arbor, MI 48105

William T. Fairgrieve, Jr.
Nutritional Svcs., ODFW
17330 SE Evelyn St.
Clackamas, OR 97017

Martin S. Fitzpatrick
Dept. of Fisheries & Wildlife
Oregon State University
Corvallis, OR 97331

Jack Fletcher
Fishing News International
A.J. Heighway Publ., Ltd.
110 Fleet St.
London EC4A 2JL, United Kingdom

Ian Forster
4466 Chaldecott St.
Vancouver, B.C.
V6S 2K1 Canada

Lee Fortier
R.W. Beck & Associates
200 Tower Bldg., 7th & Olive
Seattle, WA 98101

Dessalle Francois
Produits Trouw France
Le Pont de Pierre
Fontaine-les-Vervins
France 02140

Shunji Fujiwara
1935 'H' St., Apt. 9A
Arcata, CA 95521

O. Fukuhara
National Aquaculture Lab
Japan

Graham Gall
Dept. of Animal Science
University of California
Davis, CA 95616

Luis Maria B. Garcia
Dept. of Zoology
University of Alberta
Edmonton, Alberta
T6G 2E9 Canada

Mike Garner
Domsea Farms
4398 W. Old Belfair Hwy.
Bremerton, WA 98312

E.T. Garside
Department of Biology
Dalhousie University
Halifax, Nova Scotia
B3H 4J1 Canada

Albert Giorgi
NMFS - CZES
2725 Montlake Blvd. E.
Seattle, WA 98112

Trygve Gjedrem
Dept. of Animal Genetics
and Breeding
Agricultural University
of Norway
AS-NLH, Norway

Brian Glebe
Dept. Fisheries & Oceans
Biological Station
St. Andrews, New Brunswick, Canada

Gordon John Glova
Fisheries Research Division
Ministry of Aquaculture & Fisheries
Private Bag
Christchurch, New Zealand

Frederick William Goetz
Dept. of Biology
University of Notre Dame
Notre Dame, IN 46556-0369

Aubrey Gorbman
Zoology Dept., NJ-15
University of Washington
Seattle, WA 98195

Michael Gordon
B.C. Research
3650 Wesbrook Mall
Vancouver, B.C. V6S 2L2
Canada

Ron Gowan
Anadromous, Inc.
500 SW Madison
Corvallis, OR 97330

Dan Hallman
4209 W. Bertona
Seattle, WA 98199

David Hankin
College of Natural Resources
Humboldt State University
Arcata, CA 95521

Steve C. Harbell
Courthouse, Box 552
Montesano, WA 98563

Ron Hardy
School of Fisheries, WH-10
University of Washington
Seattle, WA 98195

Lee W. Harrell
NMFS
1624 Scandia Rd. NW
Poulsbo, WA 98370

Scott Harrington
Bldg. C-3, Rm. 124
Fishermen's Terminal
Seattle, WA 98119

Robert Harris
Washington Sea Grant
University of Washington
3716 Brooklyn NE
Seattle, WA 98105

B. Harvey
Univ. of Victoria
Victoria, B.C., Canada

Randy Hatch
Suqaumish Fisheries
P.O. Box 498
Suquamish, WA 98392

Carl Haux
Dept. of Zoophysiology
University of Goteborg
P.O. Box 250 59
S-400 31 Goteborg, Sweden

Alex Heindl
Columbia River Intertidal
Fish Commission
2705 E. Burnside, Suite 114
Portland, OR 97214

Jack Helle
NMFS, Auke Bay Lab.
P.O. Box 155
Auke Bay, AK 99821

William Hershberger
School of Fisheries, WH-10
University of Washington
Seattle, WA 98195

David A. Higgs
Dept. of Fisheries & Oceans
West Vancouver Laboratory
4160 Marine Drive
W. Vancouver, B.C. V7V 1N6
Canada

James Hill
CEDC Fisheries
250 - 36th St.
Astoria, OR 97103

Peter Hjul
Fishing News International
A.J. Heighway Publ., Ltd.
110 Fleet St.
London EC4A 2JL, United Kingdom

William L. Hobart
Marine Fisheries Review
7600 Sand Point Way NE
Bin C15700
Seattle, WA 98115

Ronald G. Hodson
UNC Sea Grant
105 - 1911 Bldg.
North Carolina State University
Raleigh, NC 27650

William Hooper
Dept. of Natural Resources, RR #6
Fredericton, New Brunswick, Canada

Glen Hopky
206-1315 - 7th Ave.
New Westminster, B.C., Canada

Charles W. (Bill) Hopley, Jr.
Washington Dept. of Fisheries
115 General Administration Bldg.
Olympia, WA 98504

Paul M. Hubbell
CA Dept. of Fish & Game
1701 Nimbus Rd., Suite B
Rancho Cordova, CA 95670

George Hunter
Fisheries & Oceans
4160 Marine Drive
W. Vancouver, B.C. V7V 1N6
Canada

Gary Ives
Suquamish Fisheries
P.O. Box 498
Suquamish, WA 98392

Robert Iwamoto
School of Fisheries, WH-10
University of Washington
Seattle, WA 98195

Bernard Jalabert
Laboratoire de Physiologie
des Poissons
Campus de Bedulieu
35042 Rennes Cedex, France

Andrea Jarvela
Washington Sea Grant
University of Washington
3716 Brooklyn NE
Seattle, WA 98105

R.J. Jenkins
New Zealand Salmon Co., Ltd.
P.O. Box 929
Dunedin, New Zealand

Charles R. Jennings
P.O. Box 5620
Elmwood Station
Berkeley, CA 94705

Christopher C. Jensen
Murray Elevators
5739 SW Chettenham Drive
Portland, OR 97201

Loren D. Jensen
15 Loveton Circle
Sparks, MD 21204

Orlay W. Johnson
NMFS-NW & Alaska Fisheries Center
2725 Montlake Blvd. E., Rm. 208
Seattle, WA 98112

William S. Johnson
Fisheries & Wildlife
Private Bag 20
Alexandra, Victoria 3714
Australia

Hirohiko Kagawa
National Institute for Basic Biology
38 Nishigonaka, Myodaiji-cho
Okazaki 444, Japan

Marcia Kamin
Seattle Aquarium
Pier 59
Seattle, WA 98101

Anne Kapuscinski
OSU Marine Science Center
Newport, OR 97365

David M. Kehoe
School of Fisheries, WH-10
University of Washington
Seattle, WA 98195

H. George Ketola
Tunison Laboratory, USFWS
28 Gracie Road
Cortland, NY 13045

Dan Kimball
USFWS
One Gateway Center, Suite 700
Newton Corner, MA 02158

Harold L. Kincaid
USFWS Fish Genetics Station
Box 700
Kearneysville, WV 25430

J.A.E. Henri Koch
University of Louvain
Zoological Institute
Naamse Straat 59
B. 3000 Leuven (Louvain), Belgium

Carl F. Koskela
Alaska Fishermen's Union
Clover Bldg., 203 W. Holly
Bellingham, WA 98225

K Koski
National Marine Fisheries Service
Auke Bay Laboratory
Auke Bay, AK 99821

B.A. Krautmann
6086 N. 6th
Fresno, CA 93710

James E. Lannan
OSU Marine Science Center
Newport, OR 97365

Mark G. LaRiviere
Northern Southeast Regional
Aquaculture Association
P.O. Box 2606
Sitka, AK 98835-2606

Sally Lawrence
Washington Sea Grant
University of Washington
3716 Brooklyn NE
Seattle, WA 98105

J.F. Leatherland
Dept. of Zoology
University of Guelph
Guelph, Ontario
N1G 2W1 Canada

David A. Leith
USFWS-Abernathy Salmon
Cultural Devel. Ctr.
1440 Abernathy Rd.
Longview, WA 98632

Eliot Lieberman
Argent Labs
14929 NE 40th
Redmond, WA 98052

N. Robin Liley
Dept. of Zoology
Univ. of British Columbia
Vancouver, B.C., Canada

Karen Pryor Lindbergh
Sea Farms Norway
44508 - 151st St. SE
North Bend, WA 98045

Seppo Lindgren
Hanka-Taimen Oy
41520 Hankasalmi,
Finland

Ron R. Loeppke
Parametrix
13020 Northup Way, Suite 8
Bellevue, WA 98005

Harold Wm. Lorz
ODFW, 4850 NW Crescent Valley Dr.
Corvallis, OR 97330

Bryan Ludwig
Fish & Wildlife Branch
218-780 Blanshard
Victoria, B.C., Canada

Phil Mackey
Mt. Lassen Trout Farm
Rt. 5, Box 36
Red Bluff, CA 96080

Colin Mackinnon
Fisheries Canada
1090 W. Pender
Vancouver, B.C. V6E 2P1
Canada

Austin Magill
National Marine Fisheries Service
3300 Whitehaven St. NW
Washington, DC 20235

Conrad Mahnken
National Marine Fisheries Service
2725 Montlake Blvd. E.
Seattle, WA 98112

John Majnarich
Biomed Research Labs
1115 E. Pike St.
Seattle, WA 98122

John H. Marsh
Columbia River Intertidal
Fish Commission
2705 E. Burnside, Suite 114
Portland, OR 97214

Laura Mason
Washington Sea Grant
University of Washington
3716 Brooklyn NE
Seattle, WA 98105

Linda May
Cockburn Bay Sea Farms
Box 159
Garden Bay, B.C.
V0N 1S0 Canada

Tom May
Cockburn Bay Sea Farms
Box 159
Garden Bay, B.C.
V0N 1S0 Canada

Carlin McAuley
Domsea Farms
4398 W. Old Belfair Hwy.
Bremerton, WA 98312

Marjorie McBride
Quinault Tribe
P.O. Box 302
Pacific Beach, WA 98571

Paul A. McCollum
Valdez Fisheries Development
Association, Inc.
P.O. Box 125
Valdez, AK 99686

Stephen D. McCormick
Woods Hole Oceanographic Institution
Woods Hole, MA 02543

Garth A. McCoy
Sea Trek
2913 - 165th SE
Bellevue, WA 98008

Ron McCollough
Mo. Conservation Dept.
Rt. 16, Box 740
LeGanon, MO 65536

Robert McKelvey
1740 Madeline Ave.
Missoula, MT 59801

Randy McLeary
Troutlodge
P.O. Box 11
McMillin, WA 98352

William McNeil
Oregon Aqua Foods
88700 Marcola Rd.
Springfield, OR 97477

C. Lynn Melby
Fisheries Research Institute, WH-10
University of Washington
Seattle, WA 98195

Jim Myers
School of Fisheries, WH-10
University of Washington
Seattle, WA 98195

Victoria Miles
Washington Sea Grant
University of Washington
3716 Brooklyn NE
Seattle, WA 98105

Mike Miller
Mt. Lassen Trout Farm
Rt. 5, Box 36
Red Bluff, CA 96080

Joan R. Mitchell
National Science Foundation
1800 'G' St. NW
Washington, DC 20037

Richard D. Moccia
International Aquaculture
Developments Company
P.O. Box 434
Erin, Ontario N0B 1T0
Canada

Kristine Moe
Journal-American
1705 - 132nd NE
Bellevue, WA 98005

Norman Moe
Oregon Aqua Foods
88700 Marcola Rd.
Springfield, OR 97477

Jeff Moore
School of Fisheries, WH-10
University of Washington
Seattle, WA 98195

Emil Morhardt
Ecological Analysis, Inc.
41A Lafayette Circle
Lafayette, CA 94549

Kelly Munkittrick
International Aquaculture
Developments Company
P.O. Box 434
Erin, Ontario N0B 1T0
Canada

Tom Murdoch
Snohomish Co. Dept.
of Public Works
Administration Bldg.
Everett, WA 98201

James J. Murphy
Dept. of Public Utilities
P.O. Box 11007
Tacoma, WA

Yoshitaka Nagahama
National Institute for Basic Biology
38 Nishigonaka, Myodaiji-cho
Okazaki 444, Japan

Masaru Nakamura
Dept. of Zoology
Faculty of Medicine
Teikyo University
Hachioji, Tokyo 192-01, Japan

Roy E. Nakatani
Fisheries Research Institute, WH-10
University of Washington
Seattle, WA 98195

Colin Nash
Kramer, Chin & Mayo
1917 First Ave.
Seattle, WA 98101

Duane A. Neitzel
Battelle NW Labs
P.O. Box 999
Richland, WA 99352

Chris Nelson
Silver Cup Trout & Salmon Feeds
118 W. 4800 S.
P.O. Box 7428
Murray, UT 84107

Rebecca J. Nelson
Dept. of Zoology, NJ-15
University of Washington
Seattle, WA 98195

Stephen G. Newman
Biomed Res. Labs., Inc.
1115 E. Pike St.
Seattle, WA 98122

Tsutomu Nishida
Center for Quantitative Science
University of Washington, HR-20
Seattle, WA 98195

Tsuneo Nishiyama
Institute of Marine Science
University of Alaska
Fairbanks, AK 99701

Richard E. Noble
Salmon/Trout Advisory Service
915 N. Quince
Olympia, WA 98506

T. Nose
National Aquaculture Lab
Japan

Terry Nosho
Washington Sea Grant
University of Washington
3716 Brooklyn NE
Seattle, WA 98105

David M. Oberbillig
Dept. of Fisheries & Wildlife
Oregon State University
Corvallis, OR 97331

John H. Oh
Weyerhaeuser Company
1454 Hayden Bridge Rd.
Springfield, OR 97477

Jeff Olsen
10135 - 35th Dr. NE
Marysville, WA 98270

Paul R. Olson
Seattle Water Department
821 Second Ave.
10th Floor, Exchange Bldg.
Seattle, WA 98104

Hiroshi Onozato
Faculty of Fisheries
Hokkaido University
Hakodate, Hokkaido, Japan

Ned A. Ostenso
Office of Sea Grant
Rm. 625, 6010 Executive Blvd.
Rockville, MD 20852

Carol Ovens
Washington Sea Grant
University of Washington
3716 Brooklyn NE
Seattle, WA 98105

John F. Palmisano
CH2M Hill
10550 - 41st Pl. NE
Seattle, WA 98125

Ned Pankhurst
Dept. of Zoology
Biological Science Centre
University of Alberta
Edmonton, Alberta
T6G 2E9 Canada

Michael H. Papst
Fisheries & Oceans
501 University Crescent
Winnipeg, Manitoba
R3T 2N6 Canada

Jim Parsons
Program in Genetics
Washington State University
Pullman, WA 99164-4350

Reynaldo Patino
Dept. Fisheries & Wildlife
Oregon State University
Corvallis, OR 97331

Merna Pelczar
4318 Clagett Parkway
University Park, MD 20782

Michael Pelczar
4318 Clagett Parkway
University Park, MD 20782

William Pennell
Malaspina College
900 - 5th St.
Nanaimo, B.C.
V9R 5S5 Canada

R.E. Peter
Dept. of Zoology
University of Alberta
Edmonton, Alberta
T6G 2E9 Canada

John Peters
Food Science & Technology, HF-05
University of Washington
Seattle, WA 98195

Patricia Peyton
Washington Sea Grant
University of Washington
3716 Brooklyn NE
Seattle, WA 98105

David E. Pflug
R.W. Beck & Associates
200 Tower Bldg., 7th & Olive
Seattle, WA 98101

Steve Phelps
USFWS Natl. Fish Health Res. Lab.
Kearneysville, WV 25430

David P. Philipp
Aquatic Biology Section
Illinois Natural History Survey
607 E. Peabody Drive
Champaign, IL 61820

Robert G. Piper
USFWS Fish Cultural Devel. Ctr.
4050 Bridger Canyon Rd.
Bozeman, MT 59715

Simon Pitney
Sevrup Fisheries
Cressy, Tasmania
Australia

Erika Plisetskaya
Dept. of Zoology, NJ-15
University of Washington
Seattle, WA 98195

Patrick Prunet
Laboratoire de Physiologie
des Poissons
Campus de Bedulieu
35042 Rennes Cedex, France

Kate Puckett
BISC, Simon Fraser U.
Burnaby, B.C. V5A 1S6
Canada

Raymond Rayhonde
Sheldon Jackson College
P.O. Box 479
Sitka, AK 99835

Frank D. Ratti
Oregon Aqua Foods
88700 Marcola Rd.
Springfield, OR 97477

Jack Rensel
Squaxin Island Tribe
West 81 Hiway 108
Shelton, WA 98584

Brian Riddell
Canadian Dept. Fisheries
Pacific Biological Station
Nanaimo, B.C. V9R 5K6, Canada

Carol B. Rideout
VA Institute of Marine Science
College of William and Mary
Gloucester Point, VA 23062

Lawrence A. Riggs
GENREC
955 Stannage Ave.
Albany, CA 94706

Ray R. Ritmer
4640 - 32nd, D-10
Bellevue, WA 98007

Steve Roberts
Wash. Dept. of Game
1421 Anne Ave.
E. Wenatchee, WA 98801

D.A. Robertson
Milnholm House, Sauchieburn
University of Stirling
Stirling, Scotland, United Kingdom

Dennis Roley
School of Fisheries, WH-10
University of Washington
Seattle, WA 98195

Peter Rombough
Pacific Biological Station
Nanaimo, B.C., Canada

Carl H. Royce-Malmgren
Zoology Dept., SLS
University of New Hampshire
Durham, NH 03820

Murray Rudd
P.O. Box 5
Madeira Park, B.C. V0N 2H0
Canada

Ernest O. Salo
Fisheries Research Institute, WH-10
University of Washington
Seattle, WA 98195

Kjartan Sandnes
Institute of Nutrition
Directorate of Fisheries
Lars Hillesgt. 26. 5013 Nygardstangen
Bergen, Norway

Richard L. Saunders
Fisheries/Oceans
Biological Station
St. Andrews, New Brunswick
E0G 2X0 Canada

Dennis L. Scarnecchia
ODFW, 17330 SE Evelyn St.
Clackamas, OR 97015

Mark J. Schneider
NW Power Planning Council
700 SW Taylor, Suite 200
Portland, OR 97205

Rich Schneider
Washington Dept. of Fisheries
115 General Administration Bldg.
Olympia, WA 98504

Carl B. Schreck
USFWS, Ore. Coop. Fish. Res. Unit
Oregon State University
Corvallis, OR 97331

Steve Schroder
Washington Dept. of Fisheries
115 General Administration Bldg.
Olympia, WA 98504

Erich H. Schuldt
Anheuser Busch
1101 Wyoming St.
St. Louis, MO 63123

Alexander P. Scott
Fisheries Lab., M.A.F.F.
Lowestoft, Suffolk
NR33 OHT United Kingdom

Tom Scribner
Confederated Tribes and Bands
of Yakima Indian Reservation
P.O. Box 151
Toppenish, WA 98948

Karl Shearer
NMFS, 2725 Montlake Blvd. E.
Seattle, WA 98112

Fred T. Shultz
Animal Breeding Consultants
Box 313
Sonoma, CA 95476

Odd Skjaeveland
Akva A/S
Oygardsvn 7
4300 Sandnes, Norway

Torolv Skretting
P.O.B. 319
4001 Stavanger, Norway

John Slind
Cockburn Bay Sea Farms
Box 159
Garden Bay, B.C.
V0N 1S0 Canada

Lewis T. Smith
Fisheries/Aquaculture/Pathology
University of Rhode Island
Kingston, RI 02881

William W. Smoker
University of Alaska
11120 Gl. Hwy.
Juneau, AK 99801

Antti Soivio
University of Helsinki
Arkadiankatu 7
00100 Helsinki, 10,
Finland

Miroslawa Sokolowska
Dept. of Zoology
University of Alberta
Edmonton, Alberta
T6G 2E9 Canada

Igor I. Solar
M.R.B.M.E.
4160 Marine Drive
West Vancouver, B.C.
V7V 1N6 Canada

Dave Somers
Tulalip Tribes
3901 Totem Beach Rd.
Marysville, WA 98270

Stacia A. Sower
Dept. of Zoology
University of New Hampshire
Durham, NH 03824

R.A.H. Sparrow
Fish & Wildlife Branch
780 Blanshard St.
Victoria, B.C. V8V 1X5, Canada

Jim Spotts
Rt. 4, Box 274
Ellensburg, WA 98926

John R.C. Springate
Fish Culture Unit
University of Aston
Birmingham B4 7ET
United Kingdom

Stefan Stippl
Am Falkeneck 1
6749 Wieslautern
West Germany

Nels V. Strandberg
Seawest Industries
100 - 2nd Ave.
Edmonds, WA 98020

Bob Suggs
1450 Telegraph Rd.
Bellingham, WA 98226

Craig V. Sullivan
School of Fisheries, WH-10
University of Washington
Seattle, WA 98195

James J. Sullivan
Sea Grant Program
University of California
La Jolla, CA 92093

Arnold M. Sutterlin
Marine Sciences Research Laboratory
Memorial University
St. John's, Newfoundland
A1C 5S7 Canada

Penny Swanson
Dept. of Zoology, NJ-15
University of Washington
Seattle, WA 98195

John Sweeney
Sweeney Enterprises
Rt. 2, Box 2452
Boerne, TX 78006

Fumio Takashima
Tokyo University of Fisheries
Minato-ku, Konan, 4-5-7
Tokyo, Japan

Wayne Thompson
International Aquaculture
Developments Company
P.O. Box 434
Erin, Ontario N0B 1T0
Canada

Gary Thorgaard
Program in Genetics & Cell Biology
Washington State University
Pullman, WA 99164-4350

John Thorpe
DAFS Freshwater Fisheries Laboratory
Pitlochry, Perthshire, Scotland
PH16 5LB United Kingdom

Roland Tomokiyo
Washington Sea Grant
University of Washington
3716 Brooklyn NE
Seattle, WA 98105

Ole J. Torrissen
Institute of Marine Research
Matre Aquaculture Station
5198 Matredal, Norway

Ed Trippel
University of Toronto
25 Harbord St.
Toronto, Ontario
M6G 2H3 Canada

Tim Tynan
Squaxin Island Tribe
West 81 Hiway 108
Shelton, WA 98584

Yngve Ulgenes
Institute of Marine Research
Matre Aquaculture Station
5198 Matredal, Norway

Timothy Unterwegner
ODFW, 17330 SE Evelyn
Clackamas, OR 97015

Fred Utter
NMFS, 2725 Montlake Blvd. E.
Seattle, WA 98112

Eric Van Den Berghe
2006 - 19th Ave. E.
Seattle, WA 98112

Glen Van Der Kraak
101-380 East 1st St.
North Vancouver, B.C.
V7L 1B5 Canada

Dan Van Dyke
224 Finley
Corvallis, OR 97331

Abe Vanderhorst
3658 Reynolds Rd.
Nanaimo, B.C.
V9T 2P4 Canada

James M. Walton
Fisheries Program
Peninsula College
Port Angeles, WA 98362

Takeshi Watanabe
Tokyo University of Fisheries
Minato-ku, Konan 4-5-7
Tokyo, Japan

R.R. Watret
B.P. Nutrition (U.K.) Ltd.
Stepfield, Witham, Essex
United Kingdom

Gary Weiner
Dept. of Fisheries & Wildlife
Oregon State University
Corvallis, OR 97331

Jason C. Wells
Valdez Fisheries Development
Association, Inc.
P.O. Box 125
Valdez, AK 99686

Alex Wertheimer
NMFS, Auke Bay Laboratory
Box 155
Auke Bay, AK 99821

Kim C. West
Dept. of Fisheries
1090 W. Pender
Vancouver, B.C.
V6E 2P1 Canada

Ulf-Peter Wichardt
Salmon Research Institute
S-810 70 Alvkarleby, Sweden

Daryl Williams
Tulalip Tribes
3901 Totem Beach Rd.
Marysville, WA 98270

Robert J. Williams
National Marine Fisheries Service
3300 Whitehaven St. NW
Washington, DC 20235

Ronald H. Williams
ODFW. 827 NW Witham Dr.
Corvallis, OR 97330

Ricardo Rodriguez Wood
Fundacion Chile
Casilla 773
Santiago, Chile

Kohei Yamauchi
Faculty of Fisheries
Hokkaido Univ., Minato-cho 3-1-1
Hakodate 041, Hokkaido, Japan

Mr. Yanno
National Aquaculture Lab
Japan

Shean-Ya Yeh
#404-1, Inst. of Oceanography
National Taiwan University
Taipei 107, Republic of China

Graham Young
Dept. of Zoology
University of California
Berkeley, CA 94720

Henry A. Zebroski
NMFS W & Alaska Fisheries Center
2725 Montlake Blvd. E.
Seattle, WA 98112